What Lies Beyond

Consciousness Science, the Paranormal, and the Post-Material Future

What People Are Saying About

What Lies Beyond

Colborn collates a wide range of fascinating information in this highly readable book that challenges the dominant materialist paradigm. He is right when he states that we are currently negotiating a 'crisis of meaninglessness', and I agree with his argument that one aspect of our path through this crisis requires a re-evaluation of the implications of psi and mystical experiences. There is a danger that the metaphysical assumptions underpinning much of modern science are at odds with what is required to negotiate the global age we are entering. Colborn addresses the issues instructively. A timely book that deserves to be widely read.
Brian L. Lancaster, PhD, Emeritus Professor of Transpersonal Psychology at Liverpool John Moores University, Alef Trust founding director, and author of *Approaches to Consciousness*

Matt Colborn makes a strong case for paying careful attention to paranormal and mystical experiences when seeking alternative worldviews to blinkered scientific materialism, worldviews that could potentially help us through difficult times ahead. What Lies Beyond is a well-balanced, informative, thoughtful discussion, and a pleasure to read.
Paul Marshall, PhD, author of *Mystical Encounters with the Natural World* and *The Shape of the Soul*, and co-editor of *Beyond Physicalism* and *Consciousness Unbound*

Well written, relevant, and from a reputable author.
Bernardo Kastrup, PhD, Executive Director of Essentia Foundation, leader of the modern renaissance of metaphysical idealism, and author of many books including *Why Materialism Is Baloney*

In this passionately argued, very accessible book, Matt Colborn directs attention to the ideal nature of science — that it is following evidence where it leads to construct an understanding of the world, not starting with a favoured conclusion and discarding all data that does not conform to expectation. The 'paranormal' phenomena that Colborn describes at length are paranormal only from the point of view of the reductionistic, materialistic assumption that consciousness is generated by the brain. These 'paranormal' phenomena are common, universal human experiences, however, that are being trivialized when dismissed so cavalierly. It is high time that our sciences faced this truth and questioned whether the materialistic understanding of consciousness is after all the most appropriate one.
James Matlock, PhD, author of *Signs of Reincarnation* and co-author of *I Saw a Light and Came Here*

What Lies Beyond

Consciousness Science, the Paranormal, and the Post-Material Future

Matt Colborn

London, UK
Washington, DC, USA

First published by Essentia Books, 2025
Essentia Books is an imprint of Collective Ink Ltd.,
Unit 11, Shepperton House, 89 Shepperton Road, London, N1 3DF
office@collectiveinkbooks.com
www.collectiveinkbooks.com
www.essentia-books.com

For distributor details and how to order, please visit the 'Ordering' section on our website.

ISBN: 978 1 80341 829 2
978 1 80341 836 0 (ebook)
Library of Congress Control Number: 2024935605

A CIP catalogue record for this book is available from the British Library.

Design: Lapiz Digital Services

UK: Printed and bound by CPI Group (UK) Ltd, Croydon, CR0 4YY
Printed in North America by CPI GPS partners

Contents

Previous Books

Non-fiction:

Pluralism and the Mind: Consciousness, Worldviews and the Limits of Science

ISBN-13: 978-1845402211

Fiction:

Emerald to Ice

ISBN-13: 979-8491513994

City in the Dusk and Other Stories

ISBN-13: 979-8457180062

For my father

Acknowledgments

Thanks first to my editor and the production team at Collective Ink. Many thanks to the following for reading and commenting on portions of the manuscript: Cal Cooper, Paul Devereux, Ed Kelly, Jim Matlock, Paul Marshall, Chris Roe, and Andreas Sommer. Additional thanks to those who provided a book endorsement, especially Les Lancaster at the Alef Trust. Special thanks to Bernardo Kastrup for recommending the manuscript to my publishers. Thanks to Elizabeth Roxburgh for her permission to cite details of the precognition study mentioned in the introduction. Thanks to Christine Simmonds-Moore for allowing me to attend her ganzfeld study so many years ago and to Matt Smith for getting me into the 2000 Perrott-Warwick conference at Cambridge. Thanks to the Society for Psychical Research for their support over the years. Thanks to the participants in my survey of psychic experience. These accounts were first published in the *Paranormal Review*. I owe a debt of gratitude to Cal Cooper and Chris Roe for a conversation that helped to determine the book's main approach. The title of Michael Grosso's book, *Smile of the Universe*, inspired the book's final metaphor. Jem Bendell's thoughts on the BuddhaAtTheGasPump podcast (8 November 2023) were crucial for framing a realistic picture of the future of transpersonal experience discussed in Chapter 10. Thanks finally to Sue, Rory, and of course to my parents for their essential moral support during the writing process. Any omissions, mistakes, or misinterpretations are of course my own.

Chapter 1

Introduction: Larger Realities?

Three curious experiences:

In November 2022, I volunteered for a dream precognition study run by a UK University. Precognition is the ability to sense or perceive a future event that 'cannot be deduced from normally known data in the present'.[1] It's thought of as very unlikely or even impossible by most practising psychologists and neuroscientists.

The instructions for participants were simple: we were to write down the content of our dreams on Monday and Tuesday night. The aim was to dream of a target that would be chosen by the computer on Wednesday. On Wednesday, 16 November, we were sent a link to a site where we were to record our dream-diaries, and afterwards get shown four possible target photographs. We were asked to choose which of the target photos most closely matched the imagery of our dreams.

Here's a record of what I dreamt on the night of 15–16 November 2022:

15–16 November.

Dream #1

Toy railways in a dark cavern, underground spaces. Lots of railway lines, set up by my father. Some of the tunnels end in my old bedroom at my old home. My father, looking at books in the long orange-painted corridor outside my old room. He is looking for something. I am very sad, nostalgic.

Dream #2

Lots of white-painted corridors and rooms. A brother and a sister team who've written about prehistoric humans. I'm drawing pictures of prehistoric humans they don't like. Christmas Day in a huge, beige room. Long tables laden with food. My family is having a Christmas meal. I don't feel very Christmassy. Then, a swimming pool. A customer is trying to help us set up some goalposts by the pool. He falls off the wave wall into the pool. I dive in and perform a rescue, towing the unconscious customer to the side of the pool.

Dream #3
Dreamt of pilgrimage through a modern town to a far place, down by the sea. A large comic-book store. Later I dreamt of two bears. One male bear tells stories to the female bear to get her to marry him.

Most of this was either garbled memory or was personal regular dream imagery. I often dream of trains; make of that what you will. I have also worked as a lifeguard. The dream of my father in our old house was a little like a distorted memory, as was the modern seaside town (I used to live in Brighton). The final dream 'snapshot', concerning two bears, was unusual.

The four possible target photos displayed on the website were accompanied by instructions to rank them in order of their resemblance to the imagery we'd already recorded. Three of the four photos had been randomly chosen by the computer. A fourth, placed in any order of viewing, was the target image. Of course, you could not know which was which from just looking at the four photographs alone. You had to guess.

The idea was to run a number of these trials with many participants and pool the results, then look for a deviation from chance in the accumulated data. If people were choosing target images by chance alone, then the researchers would expect to find that participants would choose the correct target 25%, or a quarter, of the time. To find possible evidence for dream precognition, they'd need to find a statistically significant deviation from that chance percentage (say 27% or 28% for a large study).

The first three photos were a clock, a windsurfer on a cresting wave, and a crowd. The fourth image was a baby polar bear lying on top of its mother. This was the only image that seemed close to my dream-imagery, so I ranked it on top. A week or so later, after I'd made the choice, I was sent a follow-up email. The target was indeed the polar bear photograph.

The second experience happened a few months later, while I was trekking in the Khumbu region of the Himalayas. We were visiting the village of Thame in Nepal, a stone's throw from the Tibetan border. Thame is famous for being the long-term home of Sherpa Tenzing Norgay, who summited Everest in 1953 with Edmund Hillary.[2]

Overlooking the village is a Buddhist monastery, one of the oldest in the region and the host of the annual Mani Rimdu festival. One sunny day in late March, we were led by our guide up the crooked path to the monastery. On the way, we paused beside a Buddhist shrine to look down a mountain pass fringed by snowy peaks. This pass was once part of an old trade route that led into Tibet. Further along, the path ran parallel to a mani wall, which is a lengthy stack of flat stones carved with Tibetan inscriptions of the prayer *Ohm Mani Padme Hum*. This prayer, the jewel in the lotus, was attributed to the Bodhisattva Avalokiteshvara, the embodiment of virtue and compassion.

The monastery itself is of modest size and is built around a courtyard. We were told that the interior held several *Thangkas*, religious paintings on silk appliqué that were hundreds of years old. On the way to the courtyard, we passed a small annex building with an open door. A teenage boy dressed in monk's clothes was sitting in the annex, finishing his homework. The scene looked utterly ordinary. Yet our guide told us that the boy was a *tulku*, or a reincarnation of the former abbot of the monastery. The boy had been installed as the reincarnate Rinpoche at the age of six in a ceremony on 30 July 2015.[3]

The third experience happened quite a few years ago. Here's an extract from written notes, made just afterwards:

> *As the drug [an anaesthetic] began to work, I slipped down and lost awareness and feeling in my body. Visual impressions were relatively spartan, but were mostly beige. Had experience of "white darkness" at one point, and complex patterns in dark brown against white beige, very like the side-decorations in Tibetan art…*
>
> *Reality felt gooey, or made of ice cream mixed with syrup. Feelings: at one point, I remember smiling and saying that I knew what the Buddhists were talking about. It was like I'd climbed a mountain to see a vast plain of "reality", and that it was good and light… These are metaphors; what I experienced, for fleeting moments, was ineffable.*

All three experiences in different ways seem to hint at larger realities beyond our mundane, everyday state of being. The first hints at a capacity to glimpse one's own future. The second presents a lived belief that some aspect of our consciousness might survive bodily death. The third challenges the habitual assumption of the solidity of the self, and also that we live in a reality that is impersonal, indifferent, disconnected, and dull. Together these experiences pose a challenge to a number of foundational assumptions of our rationalist, secular society. But what is their greater significance?

What This Book Is About

The large body of oft-shunned 'paranormal', 'mystical', or 'psychic' human experiences just might hold the keys to a better future. A claim like this may prompt strong reactions. One temptation might be to dismiss the proposal as dangerous pseudoscience, a distraction from the existential risks of climate breakdown and nuclear war. The opposite faith-based conclusion is also possible. Plenty of people have in the past pinned their hopes on some imminent mass-transformation of global consciousness, ushering in a New Age of peace and harmony.

Neither response is helpful. This is because it's often tough to interpret paranormal and mystical experiences, much less absorb their possible implications. Despite this, I'd still claim that the large body of what have also been termed 'exceptional human experiences' holds potential gold as well as dross.[4] 'Gold' in terms of inner capacities that might prove indispensable in the turbulent years to come.

However, they're also difficult to interpret. To return to the three examples that started the book. It's easy to think of rationalisations in order to dismiss them. Perhaps, for example, it was just a coincidence that I dreamt of bears. After all, I had a one in four chance of being able to select the target by chance alone. Also, what I'd dreamt and what the photograph showed were not identical. Perhaps I'm exhibiting a form of cognitive bias, looking for patterns that just aren't there. Confirmation bias, for example, a tendency to look for information which will confirm your already existing beliefs.[5]

As for the teenage tulku: well, religious beliefs are religious beliefs. There seems to be little or no evidence demonstrating that the boy really is the old abbot, reborn. It's just something that must be taken on faith. The reincarnation experience, from this perspective, is best seen as a social construction: a devoutly held belief, but hardly a scientific fact. However, this case does not stand alone. Across the world, a small but persistent minority of young children begin talking about past lives almost as soon as they can speak.[6] In quite a few of these cases, they can provide accurate specifics of their previous name, home, and life. They also seem to exhibit behaviours and occasional talents associated with past lives. These cases cannot be so easily dismissed.

The third experience is perhaps the easiest to rationalise. In fact, it could be taken as strong evidence that the sense of 'self' is nothing but brain function, reliant upon a functioning body. Disrupt that function with a drug, and the sense of being in a stable body is disrupted too. And the strong feelings of significance only happened because the part of the brain that produces those feelings was somehow overstimulated by the drug.

So if you wish, it's possible to dismiss these experiences rather quickly. And these mundane counter-explanations might be true, or partially true. With human beings, self-deception is always possible.

But, for me, such interpretations are not totally satisfactory. For a start, why do we feel compelled to dismiss odd or 'anomalous' experiences so quickly? Are we really being 'scientific', or is something else going on? I'm reminded of a quote from Nietzsche: 'With the unknown, one is confronted with danger, discomfort and worry; the first instinct is to abolish these painful sensations. First principle: any explanation is better than none...'.[7] With a mundane explanation, we can perhaps breathe a sigh of relief and get on with our day, our world-picture unchallenged and unchanged.

Such casual dismissals can't dispel a certain unease. This is because these experiences mark a kind of intellectual frontier. They sit on the edge of the acceptable. Speculating about possible brain malfunctions or cognitive biases is considered legitimate in mainstream psychology and neuroscience – wondering whether I'd really glimpsed the future, a higher reality, or a reincarnated person are not. It's worth thinking briefly about why that should be.

One possible reason is that psychic experiences violate or apparently violate what the philosopher C.D. Broad called 'basic limiting principles'.[8] These are common-sense principles that we tend to adopt not just in everyday life but also in science. One limiting principle is the general principle of causation. It is impossible that an 'event should begin to have any effects before it has happened'.[9] If always true, this rules out the possibility of precognitive dreams. The idea of reincarnation violates another limiting principle that's fundamental to modern neuroscience: that the mind is generated by physical processes in the brain. Broad also pointed to other limits on ways of acquiring knowledge that seem to rule out the possibility of psychic phenomena like telepathy (mind to mind communication) or clairvoyance (the mind or brain somehow gathering information from a distance without the use of the senses).

These 'limiting principles' are enough for many cognitive scientists, or scientists who study the mind, to dismiss the possibility of psychic phenomena outright. I've had colleagues assure me that we now know enough about the brain to be sure that such experiences are 'impossible'. Maybe. Yet it's also possible to claim that the dominant culture in biology and neuroscience is shaped by a number of unquestioned background assumptions that govern how these sorts of experiences are judged.

For example, the biologist Rupert Sheldrake has suggested that the worldview of scientific materialism is today dominated by what he calls ten 'dogmas'.[10] Together, these dogmas support claims that reality can be understood entirely in material or physical terms. Sheldrake's ten dogmas include the claim that reality is mechanical, or machine-like; that matter is unconscious; that nature is purposeless; that memory can be understood as material processes that do not survive death; that minds are confined to heads and are 'nothing but' brain activity; and that unexplained phenomena like telepathy are illusory.

Are We Just Mechanisms?

The first assumption, that reality is mechanical and that living things can be understood as machines, dominates contemporary biology and especially neuroscience. We will see that it significantly influences and likely limits mainstream neuroscience's attempts to understand consciousness. But it also shapes how living things as a whole are perceived.

The tendency to view living things as machines has been increasingly pervasive since the philosopher René Descartes compared animals with clockwork automata in the seventeenth century.[11] The 'clockwork' comparison changed after the twentieth-century rise of electronic devices that could respond to their environment, and especially with the arrival of digital computers.[12] Since the 1950s, the brain has often been thought of as 'processing information', in a way that is comparable to an electronic computer. But (and this is crucial to understand why things like psychic phenomena are frowned upon) the brain is not thought to be a computer with WiFi. The 'information' that living things are supposed to 'process' is thought to be restricted to that gathered by ordinary sensory means. This rules out psychic phenomena.

Fundamentally, what living things do is not to be understood in terms of minds, consciousness, or purpose but in terms of physiology, hormones, and neural circuits.[13] Cells, organs, and body systems are thought to work together in predictable ways like machines, and to produce predictable results. According to conventional biology and neuroscience, that's all there is to it.

This mechanistic approach allows a significant understanding of much animal and human behaviour. One example: the attraction of male moths to females. Male moths have antennae covered with hairs. These hairs contain smell receptors, tuned to respond to pheromone molecules (special signalling chemicals). The pheromone molecule hits the receptor and stimulates it, triggering a signal down the nerve. The signal travels to a specialised part of the moth's brain. This part in turn sends nerve signals to the wing, so that the male flies upwind.[14]

In one YouTube video, a silkworm is placed in a Y-shaped glass tube. A stick soaked with pheromone is poked into one branch of the Y. The moth shivers and then scurries rapidly towards the hormone-laced stick.[15] In another video, a researcher explains carefully how they worked out the connectome (or connections of nerve cells) governing these sorts of responses in moths. He also demonstrates a simulation that successfully mimics the functional activity of the connectome.[16] So, from a certain point of view, a previously obscure response of the moth is reduced to inner 'wiring'. Mystery is dissolved, and the robot within is revealed.

From a hard-line mechanistic perspective, living things are just robots, 'designed' by evolution. Minds are just the outputs of machine-like processes. The philosopher Daniel Dennett, who died in 2024, claimed that evolution could be understood as an algorithm (a process or set of computational rules) that 'engineers' living things. In Dennett's scheme, mind and consciousness emerged gradually through a step-by-step accumulation of mechanisms that allow survival. At the lowest levels were 'impersonal, unreflective, mindless little scrap[s] of molecular machinery'.[17] He also wrote, 'Of course our minds are our brains, and hence are ultimately just stupendously complex "machines"'.[18]

What About Consciousness?

This strongly mechanistic perspective is powerfully advocated by some consciousness researchers. In a 2017 TED talk, neuroscientist Anil Seth explains how consciousness was generated in our brains by 'the combined activity of many billions of neurons, each one a tiny biological machine'.[19] He goes on to explain how, at one time, people thought that life was more than just mechanism, but thanks to modern biology, 'people no longer think that'. So for life, there is now no need to posit the existence of an *élan vital,* or vital force. In the same way, as neuroscience proceeds,

> So as with life, so with consciousness. Once we start explaining its properties in terms of things happening inside brains and bodies, the apparently insoluble mystery of what consciousness is should start to fade away. At least that's the plan.[20]

So for Seth, life and shortly consciousness will be understood in terms of exclusively mechanistic, material processes. This is why, in his book, he terms human beings 'beast-machines'.[21] For Seth, subjective consciousness is a hallucination, generated by the brain. The self, too, is to be understood as a brain-generated illusion. He bolsters these ideas with evidence from recent neuroimaging and other studies. In the book, he also dismisses alternative approaches to consciousness for two reasons: they are 'dualistic', which means that they imagine consciousness as something 'extra' on top of materialistic processes; and they are not testable.

Superficially, Seth presents a strong, experimentally based case for human beings being 'beast-machines'. However, for me, his talk and his book raise a number of questions.

- The alternative non-materialist approaches to consciousness dismissed by Seth are responses to what is known as the mind–body problem. This problem is hundreds of years old and concerns the relationship between mind, consciousness, and the body. It is a lot harder than it looks. Is Seth right to dismiss it, or alternative solutions, completely?
- Seth's claim that we'll eventually understand consciousness in mechanistic, materialist terms in the same way that we

now understand life relies on a popular story of scientific progress. This story concerns the decline of 'religion' and 'superstition' and the rise of 'science' and 'reason'. Part of this involves the blanket rejection of things like 'vital' and 'psychic' forces and the ultimate triumph of a mechanistic understanding of life. But is this narrative true? Also, is he really justified in claiming that life is so well understood?

- Finally, Seth claims in his book that alternatives to materialism are not testable. This hints that only materialist, mechanistic explanations can be tested in science, which justifies an exclusive focus on brain and body functions. But we've already seen one attempt to find evidence for precognition, something that seriously challenges more conventional approaches to the mind and brain. Other kinds of 'exceptional human experiences' can and have been investigated and tested. Is Seth right to dismiss the challenge of these experiences?

The Enchanted Boundary

Ultimately, a purely mechanistic approach based on the assumptions of scientific materialism seems to me to result in a significantly impoverished view of human beings. One of the main reasons I think this is that, in order to make these sorts of claims, researchers like Seth and Dennett have to be very selective about the kinds of human experiences that they attend to. This is in part the outcome of training in science that restricts evidence to that found in the laboratory. It is far easier, for example, to test the brain correlations of vision than it is to discover whether we see the future, or whether reincarnation might be real, or even whether reality really is 'good and light'. But one thing I want to make clear in this book is just how extensive 'anomalous' or 'exceptional' experiences really are.

When these sorts of experiences do get acknowledged by mainstream neuroscience, they tend to get 'shrunk to fit' inside conventional worldviews. This means that the parts of the experiences that seem difficult or impossible to explain in terms of brain function or conventional psychology tend to get ignored or sidelined. For example, there is currently a great deal of attention being given to psychedelic experiences. Psychedelic experiences have many very strange features, including claimed encounters with 'entities', experiences of apparent telepathy and clairvoyance, and full-blown mystical states.[22] But these are generally interpreted as hallucinations, unusual brain functions, or even additional 'noise' in the brain.

One often gets the strong sense that there is a 'boundary of the acceptable' splitting debates over anomalous and exceptional experiences. The parapsychologist Walter Franklin Prince called this the 'enchanted boundary'.[23] To get a sense of this boundary, it's worth viewing the 2017 Breaking Convention debate between psychedelics researchers David Luke and Robin Carhart-Harris.[24] Luke is open to alternative approaches, has had entity experiences, and has also conducted research in parapsychology. Carhart-Harris tends to stick to the conventional side of the boundary, and this shows in his responses to Luke.

The 'enchanted boundary' can shift somewhat over time. Out-of-body experiences (OBEs), or the sense of being out of your body and viewing yourself or distant events, were once also shunned but are now part of mainstream research.[25] However, on the whole, researchers seem only willing to consider the aspects of the experiences that do not seem to violate the dogmas of scientific materialism. So Seth discusses OBEs as part of his argument that the self is brain-generated,[26] but he ignores potentially anomalous features of those experiences, such as reports that some people undergoing OBEs are able, on occasion, to perceive distant events accurately.[27]

Does any of this matter? One could claim that debates over fringe phenomena are only the concern of specialists. Yet the outcome of the broader disputes about consciousness surely matters a great deal for everyone. What disturbs me most about some of the predominant voices in current consciousness debates is the often implicit assumption that the *only* real way to understand human beings is in terms of what is ultimately a very crude world-picture. The cultural critic Curtis White summarises it this way: that 'we are not "free"; we are chemical expressions of our DNA and our neurons. We cannot will anything because our brains do our acting for us. We are like computers, or systems, and so is nature.'[28] This justifies the further insistence that the 'answer to all of our human problems lies in the discovery of natural laws, or that submitting to a scientific perspective is a "choiceless imperative"'.[29] As White claims, this is ideology, or scientism, and not science.

Multiple Crises

There's evidence that this predominant, scientistic worldview is causing significant problems for individuals and for society. The *Galileo Commission Report,* released in 2019, pointed to multiple crises that seem to be directly or indirectly caused by an adherence to the hard-line mechanistic, materialist world-picture.[30] These include the replication crisis in science, the ecological crisis, and the crisis of the credibility of science. For many individuals and populations, we also face crises in meaning and health.

I can illustrate the latter two crises with a personal example. For many years now, I've suffered from bouts of severe, recurrent depression. This certainly has physical aspects; a bout often follows prolonged periods of sleeplessness and stress. I seem to have a sensitivity to stress that's no doubt mediated by my brain and nervous system.

However, for me, the core of depression can best be described as a profound spiritual desolation. Often at these times I feel as if I were alone, utterly isolated in a bleak, grey landscape. Everything that I am doing feels purposeless, and the world seems drained of all meaning and is even, at times, actively hostile. It's as if I've been exiled to a dead, confining, damp, dull, grey, tomblike place.

For me, exiting this state of being means recovering a sense of meaning and aliveness in the world. Also, the recovery of a sense of deep connectedness. Accessing this means allowing space for myself, contact with nature, and practices like meditation. But the sense of meaning and connection is not something that I make up or invent. Nor is it something that conventional approaches to depression like Cognitive Behavioural Therapy (CBT) can even remotely touch. Such approaches at best alleviate the symptoms. For me, recovery is often like the recollection or rediscovery of something forgotten, a sort of *gnosis*. This shift in perception I can only call mystical.

This flies in the face of another predominant assumption: that the universe is ultimately without purpose or meaning. It challenges claims like that made by the cosmologist Steven Weinberg that 'the more the universe seems comprehensible, the more it also seems pointless'[31] and that we live in an 'overwhelmingly hostile universe'.[32] So according to the worldview of strict scientific materialism, when I'm depressed, I'm seeing the world more accurately, and when I recover, I become more deluded. I'm sure I'm not the only one to sense that there is something significantly wrong with this picture.

Harald Walach, writing in the *Galileo Commission Report*, suggests that one problem with a rigid adherence to mechanistic, materialistic approaches is that it trivialises and individualises this sort of experience. In a world that is inherently without purpose, meaning, or value, then such experiences can only ever be 'individual constructions',[33] essentially illusions of the brain generated by biology for purposes of survival. This cuts off the possibility that such states of consciousness might be letting us access deeper aspects of reality. Walach:

> The experience of meaning is itself an act of consciousness, or a gift of consciousness. In its widespread lack we see the consequence of a worldview that has relegated subjectivity to the margins or irrelevance and made consciousness a derivative of brain physiology.[34]

In other words, the kind of approach advocated by Dennett, Seth, and quite a few others acts to shut down rather than open up what might be important avenues of inquiry for individuals and for the culture as a whole. Avenues that might be essential to explore, given the multifaceted 'polycrisis' faced by our global civilisation in the twenty-first century.

Mechanistic Materialism and Power

If the worldview of scientific materialism is causing so much damage, why does it remain dominant? In part, likely because such a world-picture is supported by vested interests with considerable resources. Douglas Rushkoff, in his book on the Silicon Valley billionaires, points out that the view of 'humans as hardware' is very convenient to big tech. These 'views are entirely [...] compatible with business models that [depend] upon manipulating human beings instead of empowering them [...] exploiting them for profit rather than giving them opportunities for collective creativity'.[35]

Similar views are expressed by sociologist Brian Martin, who is interested in why alternative views of consciousness might be marginalised in a capitalist, high-tech society. He points to the pervasive influence of power and wealth on research agendas, and suggests that there might be pragmatic reasons why post-materialist views of consciousness are marginalised:

> It is plausible to argue that materialism is a useful belief system for keeping researchers and citizens oriented to serving the economy and the state rather than being 'distracted' by the pursuit of spiritually transcendent experiences. Philosophical materialism has an apparent affinity to the sort of acquisitive materialism that underpins capitalist economies.[36]

So there are reasons why hard-line, scientific materialism gets promoted favourably and other approaches get marginalised. Reasons that have little to do with science per se. The problem is that this is having a distorting effect on science. Neuroscientist Jennifer Lee offers a fairly powerful critique of the current situation. Despite money being 'thrown' at basic neuroscience research, and plenty of hype, we

> still know next to nothing about how the brain works. Moreover, though they exist, examples of emancipatory neural technologies that empower everyday people are hard to find. Instead, tepid advances in neuroscience in the coming decade will likely resemble "quick, marketable wins" in the areas of neural wearables, brain–computer interfaces, and "neuroeconomics" [...].[37]

This paints a fairly bleak picture of the immediate future, a picture that tends to be reinforced by the recent Davos session on the possibility of companies using 'mind-reading' tech to spy on their employees' very thoughts.[38] Also, with the historian Noah Yuval Harari's statement that human beings are now 'hackable animals'.[39] One has to ask who, exactly, benefits from this grim picture, and why it's being pushed so hard. The most obvious candidates are authoritarian governments and corporations.

One would also think from such pronouncements that contemporary neuroscience is approaching omniscience. And yet, at the same time, researchers have been discovering significant limits to neuroimaging technologies and AI. For example, a recent study found that machine learning failed to identify individuals suffering from major depression, based upon their brain function.[40] This indicates that despite the hype, a slavish approach to brain functioning alone might well fail to answer many of the fundamental questions with which people are plagued. This means that the search for alternative approaches assumes a fresh urgency in an era of resurgent AI and corporate-dominated neuroscience.

The Plan of This Book

This book attempts to answer two questions. First, what lies beyond the functional, mechanistic and materialistic conception of life and consciousness? Second, why does that matter today? In Chapter 2, I'll explore why the mystery of consciousness remains 'hard', and why there are reasons to doubt that the mystery will be resolved in mechanistic terms, as Seth suggests. After this, I'm going to challenge the idea that the advance of science has involved the inevitable rejection of psychic phenomena. The actual history of science suggests far more interesting and nuanced perspectives.

In Chapter 4, I'll offer a defence of studying paranormal experiences, and suggest reasons why they can't be dismissed as 'woo-woo' or 'pseudo-science'. Chapter 5 will look at the best evidence in experimental parapsychology, and Chapter 6 why the research has been so fiercely resisted by organised 'skeptics'. Following this is a look at the recent controversies over the nature of mystical states, a topic that's become more or less mainstream because of the psychedelic revival. After this, we'll look at the various expanded world-pictures that have been proposed to accommodate the full range of exceptional human experiences.

The last two chapters will change gear somewhat. Chapter 9 will critique proposed, high-tech, ultra-capitalist 'solutions' for the twenty-first-century 'polycrisis'. I will argue that purely materialist, high-tech solutions are already failing humanity in important ways, and that this failure must spur a search for alternatives. The final chapter will outline more humane alternatives and how exceptional human experiences might fit into a more benign tomorrow.

The purpose of this book is to encourage open enquiry. I hope that any reader who feels a little lost, alienated, or disturbed by the aggressive scientism and technology-worship of the current moment might take heart from the idea that there are alternative, more open approaches. These approaches reject neither science nor technology, but they do insist upon a more humanistic outlook that's sensitive to the vast, but often hidden, range of human experience.

A Note on Terminology

In this book, I'll use 'psychic phenomena' and 'psi' interchangeably as umbrella terms for telepathy, clairvoyance, precognition, and psychokinesis (psychokinesis being the alleged ability to affect the material world directly by the mind alone). Telepathy and clairvoyance are also sometimes called extrasensory perception (ESP). I will also make use of the psychologist Elisabeth Lloyd Mayer's phrase 'extraordinary knowing'.[41] 'Exceptional human experiences' (EHEs) refer to the full range of experiences that seem to lie beyond 'ordinary', waking states of consciousness. Many such experiences have also been called 'transpersonal', which means that they go beyond the personal ego of the individual. Psychic phenomena are one type of EHE; mystical experiences are another. Mystical experiences have many dimensions, but one important feature is that they seem to transform the world for the participant, and lend it significance that is not apparent in ordinary, waking states.

Chapter 2

Why Consciousness Is Still Hard

In 1998, the neuroscientist Christof Koch bet the philosopher David Chalmers that the brain mechanisms that underlie consciousness would be understood in 25 years. On 23 June 2023, at the Annual meeting of the Association of the Scientific Study of Consciousness, David Chalmers was declared the winner. In the intervening time, Chalmers stated, there had been 'a lot of progress in the field', and he felt sure that an answer would come eventually. Since 1998, he suggested, consciousness had moved from being a 'very big philosophical mystery' to one that 'we can get a partial grip on scientifically'.[1]

I was unsurprised by the outcome of the bet. 'Consciousness studies' has burgeoned in the last 20 years, becoming a vast field with many different branches. The number of different theories about brain function and conscious experience has proliferated, leading to rigorous, extensive, and ingenious scientific work. However, the dominant approaches in the field retain significant, possibly fundamental, limitations. So claims about progress remain, in a number of respects, more rhetoric than substance.

Brain-Wiring Rules?

In April 1994, the University of Arizona hosted their very first 'Science of Consciousness' conference. This had a 'circus atmosphere'.[2] The science writer John Horgan, who'd attended, later recalled the 'psychedelic vibe', and also the 'fractiousness and confusion' that had resulted from the meeting being 'a microcosm of the entire enterprise of mind-related science'.[3] This initial excitement was understandable. After decades of neglect, the topic of consciousness was coming out of the scientific cold.

A decade later, that excitement had begun to fade. The anthropologist Charles Whitehead admitted to finding the sixth Tucson 'Science of Consciousness' conference in 2004 slightly dull. Whitehead noticed that the scope of ideas and approaches had significantly narrowed. Tucson 6 had become dominated by what he termed the 'brain-wiring/information-processing faction'.[4] This hard-line materialist approach accounted for 20 out of the 24 plenary papers. 'Transpersonal psychologists, panpsychists, psychic investigators and anomaly researchers' were still present, but they were being marginalised.[5] Whitehead noticed that 'transpersonal psychology was relegated to concurrent sessions; anomalies and parapsychology reduced to a handful of posters; and the only plenary session on quantum phenomena offered few new suggestions'.[6]

The brain-wiring/information-processing approach insists that consciousness can be explained entirely in terms of physical processes in the brain and perhaps the body. It's basically materialist. The brain itself is supposed to operate somewhat like a computer. Conscious experience is therefore a brain-generated illusion, perhaps a side-effect of those physical processes.[7]

So you might think that you have free will, conscious experiences, and a self, but you are in some sense profoundly mistaken. All of these things are basically an illusion. This illusion, the story goes, can be understood in terms of known scientific principles. That is, in mechanistic, information-processing terms. The illusion probably evolved to help us survive and reproduce.

Whitehead thought that the domination of this school of thought was a problem. Although the advocates of the brain-wiring approach came up with novel ideas about consciousness, he saw the approach as 'profoundly impoverished'.[8] This was in part because Whitehead thought that the 'problem of consciousness' directly challenged 'the entire paradigmatic basis of western science'.[9] He concluded that

> The whole ethos of Tucson conferences appeared 'stuck' [...]. I believe this is because [the researchers] are not doing their job, which is to face up to the fact that materialistic approaches are self-contradictory and powerless to deal with consciousness. Backing off from the problem has led to a growing tunnel-blindness — major areas of relevant science are increasingly neglected or ignored.[10]

In the couple of decades since Tucson 6, hard-line neuroscience has become ever more dominant. In part this has been because of the influx of a significant amount of seed money. For example, in 2013, the White House announced the BRAIN (Brain Research through Advancing Innovative Neurotechnologies) initiative.[11] The beneficiaries of this new initiative included the National Institutes of Health, the National Science Foundation, and The Defense Advanced Research Projects Agency. The US Government also worked on the initiative with a number of private sector partners.

More recently, there's been a race to develop 'neural wearables' and brain–machine interfaces. Neural wearables and brain–machine interfaces, by their nature, require that you assume that the mind is like a machine with a code that can be 'hacked'. It seems fair to say that the brain-wiring/information-processing approach is in part dominant because of significant funding provided by commercial, governmental, and military interests.

But hefty funding does not automatically guarantee success. Whitehead's verdict, 20 years ago, was that '"scientists" had learned very little about the brain and nothing at all about consciousness'.[12] This might seem by now a questionable statement in light of impressive recent advances in brain-imaging technologies, machine learning, and so forth. But there are reasons for thinking that such an assessment still largely holds.

The Mind–Body Problem

In 1995, the philosopher David Chalmers introduced the 'hard problem' of consciousness. The 'easy problems', Chalmers suggested, involved the discovery of the correlations between conscious states and patterns of activity in the brain. These, he felt, could potentially be understood in materialist terms. The 'hard problem' was different. In the original paper describing the hard problem, he wrote:

> The really hard problem of consciousness is the problem of *experience*. When we think and perceive, there is a whir of information-processing, but there is also a subjective aspect. [...] There is *something it is like* to be a conscious organism. This subjective aspect is experience.[13]

Chalmers is talking about the quality of experience. These are things like the redness of a rose or the sensation and feel of a piece of music. Right now, I'm immersed in them. The washing machine is churning while I'm writing, the setting moon is white-yellow against a clear blue morning sky. If I step outside, I can smell summer scents, feel the bumps of gravel under my feet, and hear birds singing in the nearby undergrowth. I feel a kaleidoscope of different emotions, thoughts, and sensations. These 'inner messages' help make sense of the reality that unfolds about me, and my localised consciousness somehow encompasses the whole. This quality of experience is known as *qualia* or latterly, 'local states of consciousness'.[14] It is an absolutely central feature of being a living, conscious being.

You could even say that qualia constitute our whole, lived reality.

This latter fact is worth underlining because in abstract discussions, qualia can become easy to dismiss. We'll see shortly that there are even those who hold that qualia have no existence at all. The central issue highlighted by the hard problem is how these qualia, which seem utterly distinct from physical processes, might be related to the firing of neurons and other brain functions.

The hard problem is not new. It's the modern version of the centuries-old mind–body problem. The mind–body problem is a logical problem. In its more traditional form, spelled out by the philosopher Jonathan Westphal, it goes like this:

1. Mind (conscious, subjective experiences, qualia) is a nonphysical thing.
2. The body is a physical thing.
3. The mind and the body interact.
4. Physical and nonphysical things cannot interact.[15]

This mind–body problem has given rise to a range of possible solutions. Each of these solutions has significant implications for our understanding of the mind, consciousness, and reality. *Materialism*, which claims that matter is primary while mind is secondary, or illusory, or even nonexistent, we've already met. But there are other possibilities.

For instance, it could be that mind and matter are two opposite and opposed principles, things, essences, or 'stuffs' that interact. This possibility is called *dualism*. There are a number of different types of dualism. Perhaps the most famous is the substance dualism of René Descartes (1596–1650). Descartes proposed that mind and matter were two distinct essences or substances that interacted via the pineal gland in the brain.[16]

The main problem with dualism is that it remains a mystery how two opposite and opposed things might interact.

Another possibility is that mind, not matter, forms the basis of reality. This is known as *idealism*. Idealism was championed by Bishop Berkeley (1685–1753).[17] Berkeley rejected the impression that we exist within a concrete, physical world that exists independently of mind. We only perceive rocks, trees, houses, mountains, and other aspects of the physical world. We can never know them directly. This means that we've no right to talk about a rock or a mountain in itself, but only our perceptions of a rock or mountain. He proposed that nothing 'out there' in the world had any inherent existence at all. If this is true, then the question is how the world remains stable in the absence of human observers. Berkeley held that this stability was due to the all-seeing eye of God.

Dualism, materialism, and idealism are by no means the only solutions to the mind–body problem. Two more are *panpsychism* and *dual-aspect monism*. Panpsychism is the idea that consciousness is inherent in all matter, and that everything has consciousness. Dual-aspect monism proposes that the basic 'stuff' of the world is neither mind nor matter, but a neutral substance. Philosophical arguments about the mind–body problem have raged for centuries, without obvious resolution.

Today, you can find advocates for each of these alternative resolutions to the mind–body problem. The physicist Bernardo Kastrup favours idealism; the philosopher Philip Goff, panpsychism; and Jonathon Westphal, neutral monism. In the philosophy of mind, these alternatives have even enjoyed a renaissance. But despite this shift, materialist, brain-wiring/information-processing approaches still tend to get top billing. This pushes arguments about the mind–body problem in certain, perhaps predictable, directions.

Brain Function Plus Illusion

Advocates of materialism have tried resolving the mind–body problem in several ways. They tend to insist that the quality of experience, qualia, is ultimately not distinct from brain function.[18] So the sensation of the colour of a flower, or the almost numinous feeling of hearing humpback whale song, or the sweet taste of a jam-filled doughnut are in some way identical to the perceptual and brain processes that accompany those experiences. But this claim is basically a fudge. No one currently really has any idea how patterns of electrochemical activity in the brain might 'translate' into the quality of experience.

There've been numerous attempts to circumvent this problem. One is to claim that conscious experience is an 'emergent property', in the same way that the 'wetness' of water results from the mass behaviour of atoms.[19] However, this will not do. This is because consciousness is a categorically different phenomenon from the wetness of water. As the doctor and philosopher Raymond Tallis puts it, the water-consciousness

> analogy is false. The reason it does not hold up is [that] both shiny water and H_20 molecules require observation in order to be revealed as one or the other. They correspond to two different modes of observation: one by our ordinary unenhanced senses (introspecting experience, sensing water); the other by means of complex equipment and representations and interpretations that render H_20 molecules 'visible' and brain activity recordable. The two aspects of water are two appearances, two modes of experience, and this hardly applies to neural activity as electrochemical activity and experience.[20]

In other words, the concept of 'emergence' presupposes the existence of a consciousness observer in the first place. If emergence presupposes consciousness, then consciousness can't possibly be itself emergent. The argument becomes circular.

Another option for the brain-wiring/information-processing approach is illusionism. It's been suggested that 'illusion' needs to be understood not as something that does not exist, but as something that is not what it seems.[21] This is a bit like a visual illusion, where your eye is tricked into seeing straight lines as curved.

And there's evidence that visual consciousness is in some sense illusory. Experiments in change blindness show that subtle but quite significant changes can occur in a scene without people consciously noticing for quite some time. This is thought to happen in part because the brain builds up a picture of the surround using quite sparse visual information. The impression that we get of a complete scene is somehow 'filled in' by the brain. This is why Anil Seth has claimed that consciousness is a 'hallucination'.[22] Researcher Alva Noë even called visual consciousness 'The Grand Illusion'.[23]

The philosophy of 'illusionism' takes this further, stating that consciousness can somehow be entirely accounted for in illusionary terms. Illusionism's foremost proponent is the philosopher Keith Frankish. Frankish claims that we do 'not really have an inner life but illusion of rich inner life', and that 'we are introspectively aware of our sensory states but [...] this awareness is partial and distorted, leading us to misrepresent the states as having phenomenal properties [AKA qualia]'.[24]

I must admit that I find it very difficult making any sense of this statement at all. I'm quite happy to acknowledge that human experience is 'partial and distorted'. This is surely a basic limitation of the human condition. But what Frankish seems to be saying is that we think we have experiences, but we don't really have experiences. I invite the reader to think about whether this really makes any sense to them, either.

Frankish is by no means the only researcher to think in this way. The 'attention schema theory', developed by the neuroscientist Michael Graziano at Princeton University, claims that we don't have subjective experiences but instead only believe that we have subjective experiences. Graziano claims that an adamant belief in qualia is the effect of a theoretical brain mechanism called an attention schema.[25] This schema helps 'to guide, stabilize, and control selective attention, and having an attention schema can lead to an adamant belief in an ineffable something extra that we might call qualia'.[26]

This proposal also seems to me deeply flawed. There's a very significant difference between having a belief and an experience. The experience of a big red fire engine with a blaring siren roaring down the road is utterly distinct from the belief that there is a big red fire engine with a blaring siren roaring down the road. To put it another way: one can experience the fire engine without believing in it at all. As the philosopher Galen Strawson has remarked, 'it is not possible to open up a gap between appearance and reality, between what is and what seems'.[27] In other words, the strong version of illusionism just does not work.

Some Scientific Theories of Consciousness

Most mainstream scientific theories of consciousness ignore these deep problems, instead focusing on the functional parts of the brain that might support conscious experience. In 2022, the consciousness researcher Anil Seth and philosopher of mind Tim Bayne reviewed several major contenders.[28] Here we'll examine two: global workspace theory (GWT) and information integration theory (IIT).

Global workspace theory is based on the idea that conscious thinking has a limited capacity compared to the vast amount of information-processing in the brain that happens outside conscious awareness. Psychology experiments show that we're able to hold five to nine numbers in our working memory, but tend to forget more than that.[29] This limited memory 'workspace' functions like a spotlight in a theatre, with conscious events happening within that spotlight. The 'spotlight' in this metaphor brings a very small portion of the contents of unconscious processing into conscious awareness. The location of the 'spotlight' changes, depending upon where we put our attention.[30]

Integrated information theory tries to capture some key features of conscious experience in several foundational 'axioms'.[31] (An axiom is a statement or proposition that's assumed to be true). One axiom of IIT states that conscious experience exists in its own right. Another states that conscious experience is always structured and specific. A third states that conscious experience provides information to the experiencer. Further 'axioms' propose that consciousness is unitary and definite, and will always be supported by complex physical information-processing structures.

IIT claims that the consciousness in certain complex, information-processing structures can be quantified by a number, dubbed 'big Phi', (ΦMax). This and other measures express the quantity of 'integrated information' in a system like the brain. According to IIT, the amount of consciousness that a person, animal, or presumably, robot might possess depends on how much information that entity is able to 'integrate'. The human brain is assumed to integrate vast amounts of sensory information. This means that the 'big Phi' of the brain is relatively large, so we're conscious. Different mathematical values of 'big Phi' can be also used to indicate specific 'levels' of consciousness, as with people with brain damage, or who are in a coma.

Strengths and Weaknesses

Global workspace, integrated information, and other brain-function theories represent decades of careful theoretical and experimental work. Each has the potential to help us understand one or another aspect of consciousness. Global workspace theory predicts that unconscious contexts will shape conscious experience, that conscious experience will always be informative, and that 'self' will be the dominant but unconscious context for experience and actions.[32] There's at least some evidence from psychology and neuroscience for each of these predictions.

Similarly, integrated information theory at least potentially helps us to understand why correlations exist between waking consciousness and the coordinated, widespread activation of the brain. IIT also potentially shows why consciousness is associated with the activity of the brain, but not that of the spinal cord. It also helps us understand why anaesthesia causes unconsciousness.[33] IIT's predictions about the effects of sedatives have recently been tested using brain-imaging.[34]

However, these theories also have significant limitations. For a start, none of them resolve the mind–body problem. Bernard Baars, creator of the global workspace theory, is quite explicit about this. His method of testing his theory compares conscious with unconscious brain-events, but 'does not raise traditional mind–body puzzles'.[35] Information integration theory accepts conscious experience as a basic assumption or 'axiom', without further explanation. So IIT also sidesteps the mind–body problem.

Neuroscientist and parapsychologist Ed Kelly, whilst recognising the merits of IIT, also points to various weaknesses. He suggests that some of the founding assumptions or axioms made by the theory are really 'guesswork', and that the theory does not in its current form include an account of intentionality, or the capacity of conscious thoughts and experiences to be 'about' or 'directed to' something.[36] The philosopher Jonathon

Westphal finds possibly more fundamental weaknesses: it 'does not follow that if something is integrated, and this includes information, it has to manifest consciousness'.[37]

In autumn 2023, IIT theory was condemned as 'pseudoscience' in an open letter written by 124 researchers.[38] The concern was that some of the basic assumptions of IIT were not scientifically testable. Signatories included Hakwan Lau, Chris Frith, Bernard Baars and hard-line materialist philosopher Daniel Dennett, amongst others. The Affiliations section of the paper defined the group as 'A consortium of researchers concerned with the scientific status of the integrated information theory of consciousness, including its public presentation and implications.' There have been various responses to this letter. Anil Seth suggested that the 'pseudoscience' label was unhelpful, but it remains to be seen how prominent IIT will be in the aftermath of this controversy.[39]

Westphal suggests that current scientific theories of consciousness actually add very little to the solution of the mind–body problem. He also suggests that many of the current scientific theories are 'confused or unclear' exactly because they avoid the trickier problems of consciousness.[40] This very much reflects Charles Whitehead's thoughts at Tucson 6. Despite a burgeoning science of consciousness, despite very impressive advances in neuroscience, and despite some experimental confirmation of these different theories, they remain limited by the basic assumptions made by the researchers. This means that, as Whitehead suggested, the price for these advances has been a retreat from fundamental questions about the nature of consciousness, mind, and matter.

Brain and Consciousness

A common response is that such deep problems do not matter, or are a 'distraction'. It's assumed instead that continued experiments on the brain will eventually uncover all the underlying processes supporting consciousness. This will in some way 'dissolve', or render irrelevant, the mind–body problem.

Already, for many neuroscience researchers, it's obvious that mind and consciousness are simply functions of the brain. It is even seen as slightly foolish to question this assumption. One reason for this is the advent of brain-imaging technology. In his book, neuroscientist Daniel Bor even suggested that brain-imaging techniques now definitely demonstrate 'that the mind is nothing more than the brain'.[41]

This conclusion might be seen to be reinforced by the ongoing development and deployment of neurotech devices that can 'read' thoughts by translating brain function via machine learning. These new neurotechnologies do indeed seem to verge on the miraculous. For example, some recently developed technologies now allow brain function to be 'translated' into text messages.[42] This was done by an AI system translating the patterns of activity in the brain. In late 2023, an AI mind-reading cap was produced that can translate words from the brain via EEG.[43] Other studies have 'translated' images from the brain. For some, the success of such neurotechnology is proof-positive that a materialist worldview is correct.

However, there remain significant reasons to question this claim. One survey of neurotechnology was careful to make a distinction between mind, defined as mental states, and the brain correlations of those mental states.[44] The authors suggested that 'with brain interfacing technologies, neuroscience is now able to highlight some correlations between mental states and cerebral activity. There is thus some material basis for the mind.'[45] They went on to suggest that technological access to this material basis remains 'piecemeal [...]. Neural correlates remain physical imprints of the expression of the mind, but are not sufficient to be thought of as constituting the whole mind itself.' They cautioned against confusing mind per se and 'piecemeal' thoughts, as well as between 'reading minds' and decoding 'some neural imprints of thoughts'.[46] The same study also discussed deeper conceptual difficulties that limit technological mind reading, pointing out that memory simply cannot be read like a book.

Physicist Bernardo Kastrup, an advocate of an idealist solution to the mind–body problem, makes more basic critiques. Kastrup claims that these sorts of brain-imaging studies don't by any means 'prove' that materialism is true. Other possibilities remain open. Discussing one brain image extraction study, he said that

> All it establishes is that there are correlations between patterns of brain activity and inner experience, but this we already knew. Such correlations are also entirely consistent with many other metaphysics aside from materialism [...] so it doesn't privilege materialism at all.[47]

Further criticisms of simplistic brain-function theories are also possible. Christiana Westlin and Lisa Feldman Barrett recently published a paper that challenges some foundational neuroscience assumptions. These assumptions are the following:

1. *The localisation assumption:* it's assumed one part of the brain or one distinct brain-pattern is involved in creating an experience. So for example, fear can be identified with a part of the brain called the amygdala.[48]
2. *The one-to-one assumption:* that one specific pattern of brain activation will 'fit' or map onto one specific kind of psychological experience, always and for everyone.

3. *The independence assumption:* that the pattern of brain activation is all that matters in terms of experience. Once you understand the pattern of activation, you understand the experience.

Westlin and Barrett claim that these assumptions are a direct consequence of the machine metaphor, because the 'machine metaphor implies that the system can be broken down into independent, separable mechanisms, where each mechanism can be studied independently of one another'.[49] The authors think that this is wrong, and that it is better to understand living things as complex systems, with functions emerging as a result of the complex interactions of the different parts.

They also describe evidence from brain scanning studies contradicting these standard assumptions. For example, psychological categories actually have 'many-to-one' mappings in the brain.[50] This means that similar or identical psychological experiences can be associated with several different brain patterns. For example, very different patterns of brain activity were found to be correlated with different kinds of fear. So the brain activation pattern for 'pleasant fear' was distinct from 'unpleasant fear'. The authors suggest some alternative assumptions that they feel better fit the evidence. One alternative assumption is that the activity of the whole brain is important for any given experience. Another acknowledges the high complexity of the brain patterns.

A few researchers have uncovered evidence that might further challenge neuroscience. In 2017, Michael Nahm, David Rousseau, and Bruce Greyson produced a paper that reviewed evidence for potentially even more significant discrepancies between intelligence, thinking, and specific brain structures.[51] They looked at several cases of hydrocephalus, which is a condition where a child's skull gets enlarged because of an excess of fluid pressure inside. This condition can mean that children develop with mostly fluid and very little brain tissue inside their skull. A researcher named Lorber investigated one such person, who was a student of mathematics with an IQ of 140. However, according to X-rays, he had 'virtually no brain'.[52] This case was by no means isolated.

The authors also describe several other situations where the loss or damage of parts of the brain seem to have been associated with the onset of new talents, like a new musical ability where none previously existed. This suggests at the very least that the relationship between brain functioning and conscious experiences is far more complex and subtle than quite a few researchers have previously assumed.

The Enchanted Boundary Redux

To sum up: most progress that's been made in the study of consciousness has been with the 'easy problems', or with understanding patterns of brain activity that accompany conscious experience. As far as brain-function theories go, the mind–body problem remains pretty much untouched, and has even been actively avoided. The weakness of purely materialistic theories can be shown by how often proponents resort to 'illusion' to cover up fundamental limitations. To claim, for example, that people do not have experiences, but only have a false belief that they have experiences seems to me especially implausible. To repeat: an experience is not a belief.

There also seems to be limits to how far brain correlations 'match up' with specific conscious experiences. The existence of many-to-one mappings between psychological states like fear and activation patterns in the brain challenges several key tenets of neuroscience, as Wrestin and Barrett show. Lorber's hydrocephaly patients potentially push this still further. How can someone with significantly reduced quantities of brain material have an IQ of 140? This is hardly what we might call a 'neurotypical' situation. And the hydrocephaly cases do not stand alone.

Significant brain anomalies like this bring us to the edge of the known world. We arrive, in fact, at Prince's 'Enchanted Boundary'. Beyond this boundary are disputed phenomena that would seem to further defy mainstream assumptions. These include extreme psychophysiological influences such as stigmata, where devout religious believers spontaneously produce the wounds of Christ on their bodies; and calendar and calculating prodigies who have the capacity for pretty much instantaneous, complex, and specific calculations. They also include key aspects of memory and the challenge of dissociative or 'multiple personalities', some of which also exhibit extraordinary abilities. Finally, there exist significant questions about genius.[53]

This book will not be concerned with these admittedly fascinating things. Our primary concern going forward is with phenomena that seem to violate the 'basic limiting principles' discussed by the philosopher Broad.[54] Human capacities that seem to defy time and space, and call into question the tight relationship of mind, consciousness, and brain. These are variously known as psychic or 'psi' phenomena: things like telepathy, clairvoyance, precognition, and psychokinesis.

As Charles Whitehead noted, for the most part, psychic phenomena have been banished from contemporary discussions of consciousness. There are various justifications given for this: the evidence for their existence is not strong enough; they run counter to the laws of physics; they're 'irrational magic'; that they're simply impossible, 'woo-woo', or 'pseudoscience'. But the history of science shows that in the past, quite a few scientists took psychic phenomena very seriously indeed.

Chapter 3

Why the Story of Science Is Wrong

In the year 1681, a book named *Saducismus Triumphatus* ('Triumph over Sadducism') was published in England. This book, subtitled 'full and plain evidence concerning witches and apparitions', was published posthumously. The author was the clergyman and philosopher Joseph Glanvill (1636–1680). *Saducismus Triumphatus* includes an extraordinary account of happenings in the village of Tedworth, Wiltshire.

About halfway through 1661, John Mompesson of Tedworth had an encounter with an 'idle Drummer' in the neighbouring town of Ludgarshal.[1] Mompesson was a commission officer, an excise officer, and a local landowner.[2] Mompesson discovered that the drummer had a counterfeit pass, and reported him to the local Constable, 'to be further examined and punisht'. The drum was confiscated, but the drummer begged for its return. Mompesson refused, saying that he would not return the drum until he heard from Colonel Ayliff, who was named on the pass. The drum was left with the bailiff, but the following April was sent on to Mompesson's house. When the drum arrived, Mompesson was leaving on a trip to London, but on his return, his wife told him that the household had been 'much affrighted in the night'.

The house had apparently been visited by thieves, and 'like to have been broken up'. They'd also been kept awake by seemingly inexplicable sounds. Three nights later, Mompesson heard the same noise that had previously frightened his family. This was a 'very great knocking' at the doors and the exterior of the house. Mompesson toured the house with his pistols, but found no one. When he returned to bed, the thumping continued at the top of the house. These loud thumps and drums were to plague the house for the following weeks and months.

After the first month, and for the next two, the percussive noises seemed to focus on the room where the drum lay. The drumming seemed to be the worst in the first part of the night, but subsided in the small hours. The noise was soon accompanied by other strange sounds, in particular scratching noises under the children's beds and elsewhere. Then furniture began to move, and there was a 'Sulphurous smell'.

By this time, the 'Daemon Drummer' had begun to attract crowds. A church minister arrived, but in the midst of prayers the 'chairs walk't about the room of themselves, the children's shoes were hurled over their heads, and every loose thing moved about the chamber'. Later on, the drumming became responsive to questions and 'would exactly answer in Drumming any thing that was beaten or called for'. This attracted more visitors. A carnival atmosphere developed.

Glanvill himself visited and heard scratching in the bedroom of Mompesson's daughters. He searched in vain for 'any trick, contrivance, or common cause of it'. Nothing was found, and Glanvill was persuaded that 'the noise was made by some Daemon or Spirit'. The sound changed to a panting, and the 'motion it caused by this panting was so strong, that it shook the Room and Windows very sensibly. It continued thus, more than half an hour, while my friend and I stay'd in the Room, and as long after, as we were told.'

In his book *The Decline of Magic*, historian of science Michael Hunter devotes a chapter to examining the reactions to this case. The disturbances were by then attracting wider attention, not all of it complimentary. The drummer himself claimed to be responsible at his trial at the Assizes of Salisbury, and was charged with witchcraft. However, others suspected more mundane causes. It did not take long for accusations of fraud to circulate. In a letter dated 4 January 1663, Mompesson reported the arrival of some skeptical visitors who wanted to take up the floorboards after the 'spirit' declined to appear on cue.[3] The antiquarian John Aubrey reported doubts expressed by other visitors, that the drumming only occurred when the maid was in the next room. So did Samuel Pepys.[4] The Earl of Chesterfield was also critical of Glanvill's role in promoting the affair.[5] However, Hunter suggests that 'in the Tedworth case accusations of fraud were answered equally vigorously by those who believed in its verisimilitude'.[6]

In the twenty-first century, it's hard to approach the Tedworth story in an unprejudiced manner. Some might take the reports at face value, as a historical 'poltergeist' report. Others might reject any paranormal interpretation out of hand. According to some, twenty-first-century people 'know' (or should know) that magical forces and poltergeists do not exist. Therefore, the events must have been faked.

These knee-jerk responses can be a problem. The historian of science Andreas Sommer suggests that uncritical skepticism or belief can 'produce serious blind spots'.[7] He suggests that we need to set aside twenty-first-century assumptions and biases and put ourselves in the shoes of people who lived in very different, and in some ways very alien, times. This is especially important in the case of the Drummer of Tedworth, which happened at a crucial time for both culture and science.

Magic and the Royal Society

The 1680s, when *Saducismus Triumphatus* was published, were a time of transition. The first half of the seventeenth century in England had seen, along with civil war, a rash of witch trials. These trials persisted into the later seventeenth century, although they were in decline. Still, in 1682, three 'witches' were tried and executed in Exeter.[8] At the same time, there had been vigorous debates about the nature of witchcraft; the historian Rossell Hope Robbins suggested that these debates lasted for about 150 years.[9] These debates had included skeptics like Reginald Scott (1538–1599), who had written a book entitled *The Discoverie of Witchcraft*, ridiculing many of the claims.[10]

Scott's skepticism anticipated the more widespread doubts of the late seventeenth and early eighteenth centuries. The historian Roy Porter suggested that, by this time, things that had previously been blamed on witches and devils were being explained in medical terms or as natural phenomena.[11] Porter also highlighted a growing rejection of the imaginative world of witches and devils in the name of reason and progress. As early as 1658, the English writer Izaak Walton (1593–1683) had written that 'for most of the world are at present possessed with an opinion, that visions and miracles are ceased'.[12]

However, debates over witchcraft and magic took a long time to resolve. One reason was the idea that a wholesale rejection of witchcraft could also lead to 'atheism'.[13] 'Atheism' was understood very differently in the late seventeenth century. The word was a slur, and a synonym for godlessness and sin.

Some of those who opposed 'atheism' instead advocated 'supernaturalism'.[14] Glanvill's book was part of this 'supernaturalist' programme. The aim was to establish the reality of phenomena that couldn't be explained in materialistic or 'atheistic' terms. The title of the book means Triumph over Sadducism. 'Sadducism' was a reference to the sect of Jewish priests in the Bible who had no belief in the afterlife.[15] You might assume that in pushing such a 'supernaturalist' programme, Joseph Glanvill was also rejecting the new knowledge offered by science. But you'd be wrong. Glanvill was actually a member of one of the first scientific organisations in the world.

'The Royal Society of London for Improving Natural Knowledge', better known as the Royal Society, after receiving royal approval in 1663, was founded in London in 1660. It was derived from an 'invisible college' of doctors and natural philosophers.[16] The society's motto was '*Nullius in verba*', Latin for 'take nobody's word for it'. By 1665, it had published the first issue of its journal, *Philosophical Transactions*, and also Robert Hooke's book on microscopy, *Micrographia*.[17]

It might be thought that such an august scientific society would be resolutely opposed to magic and witchcraft. But once again, the situation was more complex. Michael Hunter has suggested that the private opinions of members of the Royal Society were mixed, and one might add, somewhat contradictory.[18] Some members advocated scientific investigations into magic and other 'marvellous' phenomena. These included the chemist Robert Boyle as well as Joseph Glanvill. However, other members opposed the study of magic, and were skeptical. These included the society's curator of experiments, Robert Hooke; and Henry Oldenburg, the first secretary and inaugurator of the house journal, *Philosophical Transactions*. Although even here there are complications: Hooke opposed Glanvill's investigations but was a believer in what we now call telepathy. Isaac Newton, possibly the most august member, was a dedicated alchemist but objected to the investigation of things like poltergeists.[19]

This split resulted in a stalemate over the active study of magic. The Royal Society shied away from any institutional investigations. This did not, however, stop individual members from pursuing their own interests. Hunter suggests that these individuals would follow the scientific methods laid out by the Society, but act on their own initiative. Still, the corporate indifference of the Society eventually meant that the study of 'marvellous' phenomena became excluded from the mainstream of its work. As intellectual fashions changed, and the wits of the coffee and playhouses mocked superstitious belief, the ambivalence of the society eventually hardened into outright exclusion.

Hunter concludes that the study of magic was rejected not 'for good reasons but for bad ones'.[20] What happened was that magic became targeted by humanist free-thinkers who opposed 'superstition'. From about 1700 to 1750, they waged a campaign of ridicule and skepticism against ideas that they considered superstitious. The shift from a widespread acceptance of magic to more general suspicion was the result of what Hunter calls a kind of 'cultural osmosis'.[21]

But what is meant by 'magic'? Hunter defines it as 'supposed intercourse with forces and powers above the course of nature, which it was thought possible for adepts to manipulate or control'.[22] Modern readers might think that this means that magic is essentially 'supernatural'. But historically, thinkers were more subtle than that. In medieval and early modern universities, the word 'supernatural' referred specifically to a miracle where natural laws were suspended. This was only possible, directly or indirectly, through the intervention of God.[23] So things like the phenomena of witchcraft, demons, poltergeists, and magic, were seen as only appearing supernatural. These were instead termed preternatural. Such things might be 'extraordinary' or 'marvellous', but were part of the order of nature in the way that supernatural, divinely inspired events, were not.[24]

Andreas Sommer suggests that by the eighteenth century the distinction between supernatural and preternatural had begun to blur.[25] By the time the skeptical philosopher David Hume (1711–1776) wrote his highly skeptical *Discourse on Miracles*, the distinction between 'supernatural' and 'preternatural' had been lost, and Hume makes little distinction between them. Essentially, miracles, 'marvellous', and preternatural phenomena got lumped together and were cast out of respectable society, beyond the Enchanted Boundary.

The Story

This rejection of magic is often said to be part of something called 'modernity'.[26] 'Modernity' stands for many different things. It's supposed to have begun with the scientific revolution of the sixteenth century, with the contributions of Galileo, Newton, and others. This was when nature began to be studied more systematically, with careful observation and experiment. Modernity includes a later, eighteenth-century movement that was retrospectively called 'the Enlightenment'. Enlightenment philosophers believed that you could use rational thought to improve society. This was partly inspired by the successes of natural sciences like Astronomy. If 'science' and 'reason' could be used to understand planetary orbits, then it was thought that they could also be used to build better societies.[27] The historian Stephen Toulmin called this idea 'Cosmopolis'.[28] 'Cosmopolis' was the idea that human nature and society could be understood and managed entirely rationally. This idea remains influential today.

'Modernity' turns out to be a very mixed bag, containing somewhat contradictory items. The championing of 'science' and 'reason' is only one element. It can also refer to ideas about 'progress'; the rise of technology; political revolutions, liberalism, and some forms of conservatism; socialism and communism; the advent of democratic societies; state planning; the rights of women and other minorities; large institutions governed by bureaucracies; colonialism and globalisation; steam and combustion engines; computers and brain scanners; nuclear missiles and rockets to the moon; mass production; and total war.

Modernity also involves the rejection of a number of things: the divine right of kings to govern a nation without question; the feudal system; the power of the church and religion, especially the Catholic Church; the idea of divine intervention; and, as we've already seen, magic.

This wholesale rejection of the magical, the marvellous, and the preternatural is part of a related process called 'rationalisation'. 'Rationalisation' in sociology is the replacement of traditional power structures and modes of thought by ones based on reason and rationality.[29] This was an idea put forward by the pioneering sociologist Max Weber (1864–1920) and is related to what he termed 'the disenchantment of the world'. What Weber meant by this is actually complex, but it's often been taken to mean that any form of magic, paranormal, or the supernatural must be rejected, along with traditional religious ideas.[30] Ideas like magic, 'soul', or 'spirit' must go in order to make way for totally natural explanations.

This sort of picture is championed in quite a few popular books. These include Carl Sagan's *The Demon-Haunted World: Science as a Candle in the Dark*, Steven Pinker's *Enlightenment Now*, and Richard Dawkins's *The God Delusion* and *The Magic of Reality*. Skeptical books on the paranormal like Susan Blackmore's *Dying to Live: Science and the Near-Death Experience*, James Alcock's *Parapsychology, Science or Magic?*, and James Randi's *Flim-Flam* draw upon the same modernist story. 'Enlightenment values' of science and reason are championed. Any 'belief' in magic or the paranormal is unacceptable exactly because they're conceived as the polar opposite of 'Enlightenment values'. This attitude was exemplified in Richard Dawkins's 2007 TV programme *Enemies of Reason*, which included scathing treatments of subjects like acupuncture, mediumship, and psychokinesis.

There's a problem with this story. In fact, it turns out to be significantly flawed. For example, the concept of 'the Enlightenment'. The historian of science Jason Josephson-Storm states that this only became part of the shared culture of the English-speaking world in the late 1950s.[31] English-speaking writers inherited the idea from nineteenth- and early twentieth-century Germany. Josephson-Storm notes that the idea of 'the Enlightenment' as a distinct historical period is an anachronism. People living in the eighteenth century did not think of themselves as living through 'the Enlightenment' at all. This means that writers like Steven Pinker who write books explicitly defending 'Enlightenment values' are drawing on a modern myth, and not a historical reality.

The Myth of Disenchantment

'Disenchantment' is also not what it seems. Weber's work has often been taken to mean that science and disenchantment go hand in hand. This assumption has become a common one. For example two authors of a paper pushing a materialistic interpretation of near-death experiences claim that one task of science is to 'demystify' the world.[32] 'Disenchantment' has also been taken to imply that modern European culture has no magic in it at all.[33]

These interpretations likely misread Weber's original intent. It's sometimes been assumed that Weber himself was a die-hard rationalist, a champion of 'Enlightenment' values. But this isn't accurate. In searching through the original sources, Josephson-Storm uncovered a quite different picture. Weber's letters show that he had a strong interest in mysticism. He was also a member of the occult group the 'Munich Cosmic Circle'.[34] Josephson-Storm has shown that Weber's interest in magic and enchantment happened quite late in his career, and likely coincided with a more general cultural growth of interest in magic and the occult. The conclusion is that Weber formulated his ideas about disenchantment partly as a result of his exposure to occult practices and ideas.

Josephson-Storm also challenges the idea that European culture lacks magic. A number of surveys contradict this. They show widespread psychic experiences and belief in things like telepathy, magical forces, and life after death. He concludes that 'If one looks at Europe and America through the eyes of an outsider [...] it seems hard to assert that we live in a straightforwardly disenchanted world'.[35] The rise of atheism also does not necessarily mean a rejection of magic or the paranormal; in fact, it can spur new interest. Josephson-Storm concludes that 'the death of God does not necessitate the death of magic, and if anything, secularization seems to amplify enchantment'.[36] This directly contradicts the sort of story pushed by the aforementioned promoters of 'science', 'reason', and 'Enlightenment values'.

Society for Psychical Research

The nineteenth-century rise of psychical research is one specific example of how secularisation led to 'enchantment amplification'. The Society for Psychical Research (SPR) was founded in Britain in 1882 as a response to the ongoing decline of traditional religious beliefs and the seemingly unstoppable rise of materialist science. By the second quarter of the nineteenth century, 'reluctant doubt' about traditional religious beliefs had become widespread in Europe.[37] This was part of the longer-term effects of the cultural shift described by Michael Hunter. Classical physics was ascendent, and in 1859, Charles Darwin published his *On the Origin of Species*, establishing the theory of natural selection as a foundation of modern biology. In this context, older beliefs about God began to seem obsolescent. A contemporary writer, J.A. Symonds, stated the following of this time:

> The sensation of God disappeared from me without the need for God being destroyed. But this is not merely personal history, it is the history of the age in which we live, the age of disintegration of old beliefs.[38]

The ascendency of materialism and the decline of traditional religious beliefs spurred a number of counterreactions. One was the genesis of spiritualism. This religious movement began in Hydesville, New York State, in 1848. In Hydesville, two sisters allegedly began communicating with a spirit.[39] This sparked spiritualism's rapid spread across America and its eventual arrival in England. Spiritualist mediums claimed to be conduits for communication for the dead. Spiritualist seances were also the sites of apparently extraordinary events such as raps, object movements, levitations, and the materialisation of quasi-physical spirit forms. Fraud was also rife. There were a number of scandals involving mediums who were caught creating 'phenomena' by fraudulent means.[40] However, it was by no means clear whether all such phenomena could be explained as fraudulent.

The Society for Psychical Research (SPR) was founded in 1882. Some years later, one of the first members of the society, the poet and classicist Frederic Myers (1843–1901), described the SPR's origins in the following way:

> In about the year 1873—at about the crest of perhaps the highest wave of materialism which has ever swept over these shores—it became the conviction of a small group of Cambridge friends that the deep questions thus at issue must be fought out in a way more thorough than either the champions of religion or materialism had yet suggested.[41]

This initiated an intense programme into 'whether anything could be learnt about the unseen world or no'. This was explicitly a scientific programme; Myers himself stated that, in order to proceed, 'I have appealed to Science and it is to Science I must go.'[42] The SPR's aim was to investigate, in an unprejudiced manner as possible, 'those faculties of man, real or supposed, which appear to be inexplicable on any generally recognised hypothesis'.[43]

And investigate they did. The early years of the society are widely regarded as some of its most productive. Their first major work, *Phantasms of the Living* (1886), which chronicled apparent telepathic communications between those in distress and a distant friend or relative, came in two volumes totalling 1300 cases. In this, 702 cases were examined, classified, and analysed with impressive clarity and care. Gauld observed that 'to pass from even the ablest of previous works to Phantasms of the Living is like passing from a Mediaeval bestiary or herbal to Linneus' *Systema Naturae*'.[44]

The relationship of psychical research with the rest of science was complex. It's true that the SPR had to weather its share of criticism from some eminent scientists and other critics. For example, Hermann Von Helmholtz (1821–1894) stated that neither the testimony of the Royal Society nor the evidence of his senses would allow him to 'believe in the transmission of thought from one person to another, independent of the recognised channels of sensation'.[45] Von Helmholtz was an eminent physicist and physician. He was also a campaigner for materialist perspectives in biology.

However, a roll call of some of the early SPR members reveals something that twenty-first-century readers might also find surprising. Quite a few eminent physicists and other scientists were members of the Society in its early years. Many also served as president. These included Sir William Ramsay (1852–1916), winner of the 1904 Nobel Prize in Chemistry; Charles Richet (1850–1935), who won the 1913 Nobel Prize in Physiology or Medicine; and J.J. Thompson (1856–1940), awarded the 1906 Nobel Prize in Physics.[46] Other eminent scientists who were early members included Sir William Barrett (1845–1925), the chair of physics at the Royal College of Science in Dublin and a Fellow of the Royal Society; the chemist, physicist, and inventor Sir William Crookes (1832–1919); the physicist Heinrich Hertz (1857–1894), who showed the existence of electromagnetic waves; and the physicist and mathematician Sir Oliver Lodge (1851–1940), developer of wireless telegraphy and principal of Birmingham University. The physicist Balfour Stewart (1828–1887) was a member of both the Royal Society and another SPR president, as was the astronomer F.J.M. Stratton (1881–1961).[47]

This list, which is far from exhaustive, shows at the very least that psychical research was taken far more seriously by 'hard' scientists than many popular science texts would suggest. The historian of science Richard Noakes has suggested that the influence of psychical research on late nineteenth- and early twentieth-century physics might have been far more extensive than previously suspected.[48] Noakes suggests that many of these scientists saw psychical research as an extension of the more mainstream work they were doing. There was to them no sharp division between physics and psychical research. Instead they saw the challenges posed by things like spirit-rapping, telepathy, and psychokinesis as opportunities to enlarge science.

None of this cuts any ice with critics. The skeptical writer Harry Edwards suggested that it is not the academic standing of a scientist that counts, but the validity of their arguments.[49] For Edwards, the study of magic and the paranormal has always proved to be entirely bogus. He accuses interested scientists of being fooled by their own bias and self-deception. This bias and self-deception has often, Edwards claims, been uncovered by the scrutiny of their more critical peers. And it is true that the scientists who attempted to 'enlarge science' by including paranormal phenomena like spirit-rapping, telepathy, and psychokinesis were ultimately unsuccessful. Today, we do not have a physics, biology, or neuroscience that allows us to understand telepathy or any other kind of psychic phenomena.

Still, the extent of historical scientific interest in psychic phenomena, at least on the quiet, should give us pause. For me, it suggests that the scientists who became interested should not be dismissed as isolated, deluded aberrations. And it's simply false that in every case their ideas and research were carefully examined by a skeptical but fair scientific community and found wanting. What instead often happened was that their work was conducted outside the main current of science and ended with their retirement or death. This meant that many of the questions that led them to study paranormal or psychic phenomena remained unanswered.

What also happened was that intellectual and cultural patterns shifted. By the early twentieth century, psychical research was being marginalised. In part this was due to the rise of behaviourism, a new form of psychology that saw animals and humans as blank slates whose nature could be entirely accounted for by chains of learned behaviour. 'Consciousness' itself was put on the rubbish dump of discredited ideas. One of the founders of behaviourism, John B. Watson (1878–1958) stated that the time had come 'when psychology must discard all reference to consciousness [...]. Its sole task is the prediction and control of behaviour, and introspection can form no part of its method.'[50] If ordinary mind and consciousness was seen as dubious, then psychic phenomena had no chance.

The Lessons of History

To sum up:

Today, popular books often promote a simple story about the advance of science. The story goes that magic, paranormal, occult, or vital forces had to be dismissed in order for science to progress. This story also holds that dismissing these things allowed a mechanistic view of life and mind to triumph. This is the narrative underpinning Anil Seth's 2017 TED talk on consciousness. In the talk, it's claimed that a belief in consciousness as something special or 'magical' will fall away once the underlying 'mechanisms' are understood. This is a direct appeal to 'disenchantment'.

Sometimes it's also claimed that magic, paranormal, occult, or vital forces were tested via the 'scientific method' and found wanting, which also justifies their exclusion from science. At other times, psychic phenomena are summarily dismissed as 'pseudoscience' because they're claimed to be impossible in terms of current physics, or because of what we now know about brain function.

The historical record tells a different, more nuanced story. Michael Hunter and others' work on the early history of the Royal Society shows that its members were split on the topic of magic and 'preternatural' phenomena. As a result, for diplomatic reasons, the topic was excluded from the Society's main work. Despite this, individuals in the society continued to pursue work on phenomena like poltergeists and second sight. But an ongoing cultural shift meant that, eventually, 'preternatural' phenomena were consigned to Gothic fiction or became ridiculed. So 'magic' declined from 1700 to 1750.

Later on, in the nineteenth century, the decline in traditional religious belief in Europe and America not only spurred novel religious revivals but also the investigation of psychic phenomena. These investigations involved a range of top researchers of the day. These researchers often worked outside the 'mainstream' of science, but they hoped to extend science to include things like telepathy, psychokinesis, and table-rapping. These efforts eventually fell into eclipse as the researchers died and the intellectual climate shifted in a materialistic direction. Despite this, it's incorrect to call modern industrial consumer societies 'disenchanted'. Some elite interest in psychic phenomena has persisted, and belief in the paranormal remains widespread.

I'd also return to Michael Hunter's view that the Royal Society shied away from the investigation of 'preternatural' phenomena not for good reasons, but for bad ones. The reason this was a bad move is simple. Dismissing a significant body of human experience by ridicule is not science at all. It's a social, political strategy. This means that this significant body of human experience remains available for investigation, no matter how exotic, 'marvellous', or unlikely some of the reports might seem. It's surely very unscientific to turn one's back on a potential source of information that might teach us one or two novel things about consciousness and the mind.

Finally, if the 'disenchantment' of Europe and America is a myth, then is it still accurate to claim that scientific materialism is predominant? I think so, for the following reason. Despite the popular persistence of paranormal and religious ideas, it's still true that large institutions in industrial consumer countries remain tacitly materialist. This includes governments, academia, and the medical and legal professions. This means that the assumptions underpinning these institutions will include scientific materialism by default. In addition, vested interests in big business also mostly accept a materialistic outlook.

Up to a point, this dominance narrows the field of vision of both science and the humanities. It creates the cultural 'blind spots' mentioned by Andreas Sommer. These 'blind spots' mean that within most biology laboratories, life will only be viewed in certain ways. Within neuroscience laboratories, consciousness will only be viewed in certain ways. Research will also be narrowed by the requirements of business and a need for fast profit; so life, through a commercial lens, becomes something to be patented, manipulated, and created for profit. Consciousness becomes 'nothing but' brain mechanisms, with those mechanisms accessed, monitored, and controlled by large corporations. This is the implication behind claims that life, and shortly consciousness, will be comprehensively understood in mechanistic ways.

For me, the lesson of history is that things do not have to be so narrow; that it's quite possible for scientific research to take other directions; that it's possible to pay attention to extraordinary human experience; and that it's possible, if not easy, to seek expanded ways of viewing the world, life and consciousness, and to pursue this in a thoroughly scientific spirit.

This lesson becomes salient when we begin to think about contemporary paranormal and psychic experience. In the following chapter we'll see that Izaak Walton was wrong and that 'visions and miracles' have not ceased. We'll see that experiences of apparent telepathy, clairvoyance, precognition, and the rest remain a subtle but persistent part of human experience. This is unsurprising because the story of the banishment of 'magic' is just that, a story.

Chapter 4

The Significance of Paranormal Experience

In 1977, the explorers and film-makers Lawrence and Lorne Blair sought out some of the last nomadic Punan Dyak people of Borneo.[1] They travelled past coastal logging towns, heading for the island's rugged interior. They rode boats westwards along the Mahakam river and up its tributaries, past scattered logging camps. North of this region was a white patch on their international aviator's chart, the white signifying unmapped territory. They ended up travelling in the rainy season, hoping that the rivers would be more navigable when the water was at maximum height. Unfortunately, this meant encountering large swarms of mosquitoes.

In a village called Tabang they were told of a guide, an old Punan rhino hunter named Bereyo. Seeking Bereyo, they trekked into the jungle with a party of Dyak people. On the sixth day, they found him living in a residence on the Belinau river. He agreed to help with their search, which he said might take 25 days.

This was a tough journey over treacherous terrain. The brothers noticed that the forest was composed of 'ecological islands' of great diversity and enjoyed beautiful views of waterfalls and tropical birds. There were plenty of hazards, mosquitos, leeches, and snakes. Eating was also tough. After almost a month's worth of travel, their supplies had begun to dwindle and they relied on the Dyak hunter's hunting skills to catch monkeys for protein.

However, their guide seemed to know where he was going, despite not having visited that region for over 20 years. When Lawrence quizzed Bereyo on this unerring sense of direction, he replied:

> We Punans know that we have two souls. There's the physical, emotional soul [...] and the "dream wanderer". In sleep and special trance, the dream wanderer travels, sees with different eyes, sees wild pathways or lost people.[2]

Then for 8 days there was a torrential highland rainstorm and the river burst its banks. The rain sometimes fell as 'solid lumps', and the expedition was immobilised. The brothers' fellow Dyak people didn't seem overly concerned. It turned out that Bereyo had dreamt that the people they sought were sheltering in some abandoned longhouses further down the river. Their companions were so confident that Bereyo was right that they had begun to groom themselves for the forest women they felt sure that they would encounter. As soon as the rains slackened, Bereyo sent out a six-person scouting party. After another few days, the party confirmed that the nomadic Punan people were exactly where Bereyo said that they'd be.

Judging Indigenous Accounts

How should we judge this account? Think about how Bereyo's apparent ability to 'know' things from afar presents itself. The account makes it fairly clear that 'dream wandering' was an integral and accepted part of the everyday life of the Dyak people. They accepted the ability accurately to dream of distant places and people in the same way that people in industrial consumer cultures accept that mobile phones can connect them to people across the world. This acceptance is shown by their preparations to meet the forest people. The ability serves a very practical purpose, allowing them to locate the nomadic Punan people. In this sense, dream wandering 'fits in' fairly seamlessly with the everyday. It's not perceived as 'odd' or 'anomalous', although it may well be 'psychic' in some sense.

Finally, the ability is part of a familiar, universal psychological phenomenon: dreaming.

This continuity with everyday life and with already familiar – if imperfectly understood – mental processes is important. We might expect any genuine but unknown mental process to manifest itself within known psychological processes or abilities. This does seem to be the case. Apparently 'seeing' the future or 'knowing' about distant events or people is often 'mixed up' with dreams, or states of reverie, or even with everyday, rational states of consciousness. There have even been suggestions that psi information seems to be recalled like 'ordinary' memory.[3]

But is there any need to invoke an unknown process to explain dream wandering? Strictly speaking, and taking the account to be an accurate reporting of events, there remain possible conventional psychological explanations for Bereyo's apparent 'extraordinary knowing'. There's no doubt that forest-dwelling peoples have a far greater sensitivity to their environment than urban dwellers. Perhaps, for example, Bereyo's brain was processing subconscious but quite ordinary environmental cues picked up during the journey and presenting the results of ultimately lucky guesses to him in the form of a dream. On one level, this sort of speculation is legitimate and not unreasonable.

But it also amounts to a refusal to take what Bereyo says seriously. Bereyo claimed that his 'dream wanderer' left his physical body, travelled far, and returned with useful information about 'wild pathways and lost people'. From Bereyo's point of view, there were very good reasons to think that this was true. For him and the other Dyak, 'dream wandering' worked. It was a reliable means of gaining information. However, for people living in techno-scientific cultures, this sort of claim is very easy to reject almost without thinking.

This sort of unthinking rejection has had negative consequences. In the early 1980s, the anthropologist Jeremy

Narby lived among the Asháninca people in the Peruvian Amazon. The Asháninca made use of a hallucinogenic brew, ayahuasca, to gain knowledge from the forest. This included knowledge about the use of medicinal plants and techniques for forest agriculture. Some of this knowledge turned out to be highly sophisticated, effective, and subtle. Narby was told that 'ayahuasca is the television of the forest. You can see images and learn things.'[4] Narby later took ayahuasca and had some very powerful visions himself.

Narby reports that for a good portion of the twentieth century, anthropologists had failed to grasp what their informants were telling them about such shamanic practices. Some of this ignorance was grounded in the racist assumption that indigenous peoples were 'primitive' and therefore ignorant. In the 1980s, this sort of assumption remained an active force in the Peruvian Amazon. Narby's own investigations of 'Asháninca resource use' were linked to the activities of international development agencies who were then pouring millions of dollars into the development of Peru. The Asháninca and other indigenous people were looked upon as children. The 'development' of the Peruvian Amazon consisted of confiscating indigenous territories and turning over the land to developers who would chop down the jungle, replacing it with cattle pastures. Narby says that experts 'justified these colonisation and deforestation projects by saying that Indians didn't know how to use their lands rationally'.[5]

Very often in the Global North, debates about paranormal and anomalous experiences are treated as light entertainment. They're often 'silly season' topics, something to be discussed when there's little serious news to report. There's often a jokey or condescending tone to this discussion. People who 'believe', it's often assumed, must be irrational or in some way gullible.

But in the case of the Punan and Asháninca people, 'anomalous' experiences were anything but anomalous. They were part of their knowledge system that helped them live in the world. It is true that these knowledge systems look very strange from the point of view of people in the Global North. However, they functioned well in their local context. Narby and other anthropologists found good evidence that the Asháninca people were able to derive sophisticated knowledge from their knowledge system. Bereyo's 'dream-wandering', however we might understand it, also furnished him with practical knowledge.

So here's one reason to take psychic experiences seriously. They often formed an integral part of traditional and indigenous societies. The same traditional and indigenous societies that for the last few centuries have often found themselves overwhelmed or even destroyed by colonialism and war.

What About Modern Societies?

The situation in countries like the UK and the US is very different. These countries support what's been called an industrial consumer way of life. In other words, they are kinds of society geared towards the production and consumption of goods and services. Scientific materialism in a sense forms the official belief system of industrial consumer societies, which means that psychic phenomena are not supposed to exist.[6] Paranormal experience is often marginalised.[7]

Despite this marginalisation, psychic experience remains a relatively common, if sporadic, feature of human life. A 2009 survey of the UK found that 39% of people had reported at least one paranormal experience.[8] These include ghost experiences and also apparently telepathic, precognitive, or other 'extrasensory perception'-type experiences. However, these events are very often not really integrated into people's lives.[9] They are often seen as exceptional, perhaps intrusive — certainly not part of the everyday fabric of existence.

Despite this, some psychic experiences in the UK have a family resemblance to the 'dream wanderings' of the Dyak.

The underlying phenomenon, of 'extraordinary knowing', is consistent. In the 2000s, I conducted a small survey of psychic experience. One respondent wrote: 'My mother woke up in the middle of the night and thought I had had a car accident. I was supposed to be asleep with friends. One minute later the police called to tell her I had had a car accident.' A 24-year-old man recounted an experience as a teenager:

> *When I was sixteen, my sister and I were sitting in our lounge watching TV while she was waiting for her boyfriend to turn up. He was running a little later than normal, and she was beginning to worry about him. All of a sudden, I said to her, "He's probably crashed his car and is lying in a ditch somewhere." The instant I finished my sentence, the telephone rang, and it was her boyfriend, and she said that he said, "I've crashed and I'm lying in a ditch somewhere." Almost the exact words that I'd said. There was no way I "knew" that he had just crashed, and no way at all for me to know that he wouldn't know where he was. My sister was freaked out... and wouldn't speak to me for a few hours.*

These sorts of experiences are tantalising but inconclusive. Was this just a coincidence? It's difficult to tell. Others reported apparently premonitory dreams.

> *I dreamed around the age of 13 that my mother had used Bovril as a starch for soup. She had never done this before, and I thought it unusual. The next day she was making soup, and putting Bovril in it. I questioned her about this. She said it was the first time she had ever used it for the purpose of making soup.*

This is again curious, but it was recorded many years after the fact, and could also be put down to coincidence. It's quite possible that quite a few cases of apparent premonitions, or 'sensing the future', can be explained in ordinary psychological terms. A team led by the psychologist Caroline Watt found that people recalling precognitive dream experiences tended to be selective in what they recalled and that they showed a tendency to find correspondences between events that were really unrelated.[10]

However, some of the accounts of apparent precognition seem to defy such straightforward analysis. Here is one from my survey, reported by a 45-year-old woman:

> *During the early part of 1996, I started getting flashes of visions of a building. These happened during the daytime, whilst I was going about my work, either whilst I was sitting at my desk or when I was driving my car. These continued periodically over a few weeks. Each time I'd get a bit more information and see a bit more of the building and also a man walking through it. The flashes became more regular and the images clearer, together with an urge to 'do something' or 'tell someone'.*
>
> *I could ignore this no longer, so I asked a colleague to listen to me and I told her that I thought there was going to be "another Hungerford disaster". I could "see" a man walking through a building dressed in camouflage gear and looking like "Rambo" with a number of guns and bullet belts crossed over his chest.*
>
> *I told her I could see him walking through the building, which had a lot of desks in it, and he was shooting people indiscriminately. There was blood and bodies everywhere.*
>
> *One week after I told my story to my colleague, the news carried details of the Dunblane disaster. When I saw the news report, the pictures of the school were exactly what I had seen in my head. I cannot tell you how disturbing this experience was.*

The Dunblane massacre took place in Scotland on 13 March 1996.[11] It took place at the Dunblane Primary School, when Thomas Hamilton shot dead 16 pupils and one teacher, injuring 15 others, before shooting himself. It was the deadliest mass shooting in UK history. The Hungerford massacre, which the witness also mentioned, was a spree shooting in Hungerford, England on 19 August 1987.[12]

This account seems to cast doubt upon the chance-coincidence theory. Firstly, mass shootings are exceptional occurrences in the UK. Secondly, the vision had persistence over a few weeks and was fairly specific. Third, the witness told a co-worker before the event had occurred. Finally, the building she'd 'seen' in her vision seemed to match with the photograph of the school building on the news.

It is true that some of the specific details of the visions are at variance with the facts. For example, Hamilton was dressed in black, not 'camouflage'.[13] He also did not look like 'Rambo' and did not have bullet belts crossed across his chest. This contrasts with the witness's claim that the photograph of the school she eventually saw on the news matched up with the one in her vision. These details are consistent with other cases of premonition. Witnesses seem to encounter their own future, not the future in general. The camouflaged figure in her vision could also be a symbolic representation of the shooter, rather than a literal image. This would explain the discrepancy.

A second striking premonition account dates from December 2004. This was reported by a 44-year-old woman. I was able to confirm this account in January 2005 with a phone call, and I also got corroboration from her partner.

> *I have had a very recent experience, several days before the Christmas just gone. My husband, myself, and a dear friend of ours were seated in the kitchen. We were talking about the ice caps melting, global warming, and more to the point, rising*

water levels and how it would affect us. I became very agitated and said that we really had to think about moving because very soon we would be underwater and that there was going to be a massive flood. I was quite adamant in my feelings that something disastrous was going to happen and began feeling this high state of anxiety throughout the night. The feeling has haunted me, always niggling at the back of my mind, and then we saw the news on boxing day of the Tsunami. I realise that this disaster took place in the Southern Hemisphere, but I couldn't believe how I'd been feeling for a few days before it happened and the animated discussions I was having regarding flooding.

On 26 December 2004 at 00:58 UTC, the 'Boxing Day Tsunami' struck the coasts of India, Indonesia, and Thailand.[14] Witnesses spoke of a wall of water, with waves up to 30 m (100 ft) in height. The tsunami was triggered by a major earthquake with a magnitude of 9.1 to 9.3, with an epicentre off Sumatra, Indonesia. A total of 227,898 people died.

What is striking about this account is the strength of emotion associated with mostly symbolic visions of flooding. The persistence of this feeling until the confirmation of the tsunami via the TV is also worth noting. The account is somewhat strengthened by corroboration.

None of these accounts amount to 'proof' of paranormal or psychic processes. But I hope that they arouse curiosity. I'd also point to the nuance of the cases. A generic dismissal citing 'coincidence' or 'cognitive errors' just won't do, and won't be satisfying for the witness. It is true that these cases are incomplete, and that a researcher must be on the lookout for distortions of memory, errors of perception, exaggeration, etc. But a blanket dismissal would not be a helpful response. Unfortunately, in official culture, this is too often what witnesses get.

Hidden Events

2018, the UK talk show *Lorraine*:

The science populariser Brian Cox assured viewers that ghosts couldn't possibly be real. 'Science', Cox stated, now knew enough to be sure that ghosts couldn't possibly exist. Brian Cox is a physicist and a former keyboardist for the dance group D:Ream. Cox was on the show talking about scientific discovery and his new documentary. The broadcast date was Hallowe'en. Cox drew a very firm line between what he saw as legitimate questions and what he didn't. He explained that he was 'very skeptical' about things like ghosts, and that it was just about people's 'perceptions' and 'processing stuff in our remarkable brains'. So when 'something strange happens, we attach a great deal of weight to those things' but this is mistaken. We apparently 'understand the science very well', presumably well enough to rule out the mere possibility of paranormal phenomena.[15]

The sociologist Ron Westrum has pointed to the problem of what he called 'hidden events'. He defines this as 'some unusual phenomenon that is widely experienced, but is counterintuitive. Because it doesn't make sense, the persons who experience it shrink from treating it as real.'[16] His example was 'battered child syndrome' in the 1950s. At the time, quite a few physicians were called upon to treat children who regularly turned up at surgeries with bruises or broken bones. The doctors often explained these injuries in terms of falls and other kinds of accidents. This was in part because childhood abuse was a painful, taboo topic. People, including medical professionals, did not want to know it was happening.

This seems somewhat comparable to the official treatment of paranormal experience in contemporary culture. 'Hidden events' are often far more widespread than people realise, but witnesses are often unwilling to report them. Westrum also said that experts 'are often as ignorant as everyone else, but do not know they are ignorant'.[17] This ignorance often manifests itself when experts make public statements about paranormal phenomena. Sometimes, psychic experiences are simply dismissed. However, very often the dismissal is subtler. When paranormal or 'anomalous' experiences do get acknowledged by experts, they're rapidly dismissed in mundane terms, mostly as mistaken perceptions or hallucinations.

In 2023, the *Daily Mail*'s psychiatrist Dr Michael Mosley wrote an article describing his wife's 'ghost' experience in their bedroom.[18] Mosley explained that his wife was suffering from sleep paralysis, a curious state between waking and sleep. With sleep paralysis, you can't move, as your muscles are still waking up. Sleep paralysis is often associated with a semi-dream state, so you are prone to hallucinations. It's a phenomenon that's fairly well understood in terms of brain and body function. Mosley assures us that neither he nor his wife believes in ghosts 'for a moment'. So the possibility of paranormal experience is instantaneously debunked.

Sleep paralysis is a likely explanation for some ghost experiences. In his book *Ghosts over Britain*, ghost hunter Peter Moss described the experience of Joy McKenna in 1971. Joy awoke with a strong sense of presence in her room and had an 'awful sense of overpowering fear', as her body was in a state of 'rigid catalepsy', as something with immense weight leapt on her and 'began physically to pummel her shoulders, arms and chest' like a 'demented beast'.[19] Joy prayed to God, the pressure lifted, and she could move again. Moss concluded that she'd been visited by some 'impersonal, illogical, uncontrolled and wholly evil supernatural force'.[20] Although dramatic and frightening, this encounter reads very like sleep paralysis.

Other ghost sightings can probably be explained as hallucinations. Here are a couple from my survey:

In our previous house, I saw a man sitting down on one of the kitchen chairs, looking at me. The vision only lasted a fraction of a second. This happened on a couple of occasions that day.

And this

When I had recently moved into a flat, I was sitting up late one night and was aware of a figure sitting on the settee. The figure was of a man, and it was as if I was seeing an image of someone who was far away, as the expression on his face was unseeing, and the figure did not move. The flat had recently been renovated, and I was one of the first new tenants. It may simply have been that I was over tired and had nodded off without realising, but I felt awake at the time.

Hallucinations are actually very common. A recent large online survey found that 29.5% of respondents had experienced auditory, and 21.5% visual hallucinations.[21] These hallucinations are also often associated with sleep states. They're interesting but not really challenging to conventional understandings of human psychology.

But the central issue is not whether some kinds of anomalous experience can be entirely explained in conventional terms. This is because hidden events refer to causes and not to the visible evidence of those causes. So in the case of childhood battery, experts regularly witnessed bruises and broken bones, but rationalised them in terms that avoided the real, distressing cause. Assuming that some psychic capacities are real, and that at least some of the accounts we have describe them, then dismissing the experiences as entirely the result of 'processing stuff in our remarkable brains' (i.e., errors) would be a mistake. The 'hidden event' we're looking for is not the report of an anomalous experience per se but the unknown process that might cause elements of those experiences.

Telepathy: A Hidden Event?

I've already said that we should expect any unknown processes to be entangled with known ones. The key question is whether any of these experiences seem to have persistent features that point to those unknown processes. The founders of the Society for Psychical Research certainly thought so. One of their first major surveys, *Phantasms of the Living,* was a compilation of case study evidence that they thought pointed to the existence of telepathy. (Frederic Myers, one of the authors of the book, actually coined the word 'telepathy').

The authors, Edward Gurney, Myers, and Frank Podmore, were well aware of the association between psychic experiences and the sleep cycle. A chapter in the book deals with apparently telepathic dreams. Another looks at 'borderland' cases that occur between waking and sleep. This is a state of consciousness known as hypnagogia. As you drift off to sleep, dreamlike imagery can intrude into your consciousness, making you hallucinate. This phenomenon can explain quite a few 'bedroom encounter' - type cases of ghosts and alleged alien abductions.

The authors of *Phantasms* were well aware that hypnagogia was a hallucination-prone time. They wrote that hypnagogia, or the border between waking and sleep 'happens to be favourable to sensory hallucinations in general'.[22] However, they were looking for more than hallucinations. They were looking for what they termed *veridical hallucinations,* or hallucinations that seem to provide the experiencer with extraordinary knowledge, and 'phantasms that coincide with reality'.[23]

The following is one strong intuitive case collected by Gurney, Podmore, and Myers:

> My grandmother, a lady considerably over 70 years of age, resided with my parents, and I was [at the time of the occurrence] staying at a place about 4 miles away from

> home. Everybody at home was, to all appearances, in good health and had been so for a long time, and on that particular night I went to bed and fell asleep, without at all divining what was soon to happen.
>
> I have no remembrance of having dreamt, and all I know is that, after having slept for a couple of hours, I woke with full conviction that my grandmother had been taken suddenly ill, that a messenger was on his way to fetch me, and that I should not reach home before it was all over. A moment or two later I got up, lit a candle, looked at my watch, dressed, and waited at the window in the full belief that my grandmother was then dead and that I should have to go presently; and as I expected, so it was, the messenger arriving just as I was ready to return with him, and the death happening, as it proved afterwards, at the very moment I had looked at my watch.[24]

For Gurney, Podmore, and Myers, experiences like this were strongly suggestive of telepathic communication. In the 1300-page, two-volume *Phantasms of the Living,* they amassed 702 cases. These cases also included 'crisis apparitions', a particular type of 'veridical hallucination' that often seemed to coincide with accident or death. Here is another 'borderland' case that occurred between waking and sleep:

> My brother Henry died in Exeter in July, 1855. I was then on a voyage home [from Shanghai]. I was very ill at the time. We were within one or two degrees of the line, fearfully hot. I had been in bed about a couple of hours and was wide awake, when I saw [...] a vision [...] of Henry. I immediately called out to my fellow passenger: "Frank, my brother Henry is dead, I have just seen him."[25]

The witness's brother died according to the register of deaths on 19 July 1855. The authors of *Phantasms* estimated that the vision had likely preceded the death by about 4 hours.

One objection might be that, although interesting, such narratives are really 'only anecdotes' and so are not admissible as scientific evidence. Certainly, accounts and case studies have their weaknesses. Chris Roe points out that they can suffer from various types of bias; that the events described are often unrepeatable; that they rely on recall, which is often imperfect or distorted; that they occur in uncontrolled conditions; and that they tend to be less convincing to the scientific community than experimental work.[26]

However, none of these are reasons for dismissing spontaneous case reports outright. Roe also points to the advantages of case studies: that case collections are often rich in the kind of detail that theories would need to explain, and that informative patterns can potentially be found in the masses of data in collections. He suggests that case studies can explore phenomena that can't be brought into the lab; that they can alert researchers about unexpected or unsought phenomena. Finally, they can complement and inform laboratory experiments.

Summing It Up

There are several reasons for taking psychic experience seriously. They've been a significant and persistent part of human experience since records began. They've been an integral part of many traditional cultures, at least until they were overwhelmed by the global spread of industrial consumer societies. These experiences quite often have a significant emotional impact on the experiencer. Even if they are regarded as trivial by the dominant culture, they are often not trivial for the witness. Finally, many of the best spontaneous cases quite forcibly suggest the occurrence of 'hidden events', specifically, mental processes unknown to science. These surely need investigation.

One common critical response is that these experiences are taken seriously, but that they can be entirely understood in terms of perhaps unusual but ultimately mundane psychology. There's no denying that this approach has strengths. We've seen that psychological events like hallucinations and sleep paralysis can account for some of these experiences. Cognitive biases and errors, where people see meaningful patterns in patternless events, are also significant. However, I'd concur with Chris Roe's view that 'this approach adopted in isolation seems akin to acknowledging that some people who claim to be ill are prone to hypochondria'.[27] Very often, conventional explanations get trotted out in order to dismiss these experiences as trivial. Any aspect of an experience that might challenge conventional thinking gets rationalised or ignored.

For me, the detail of some of these accounts sets them apart from cognitive errors, hallucinations, or other neural misfires. The stronger precognitive cases in my survey seemed to me to resist mundane explanation. Part of this was their persistence and emotional impact. The accounts gathered by Gurney, Myers, and Podmore also contained puzzling details. The witness description that 'I woke with full conviction that my grandmother had been taken suddenly ill, that a messenger was on his way to fetch me, and that I should not reach home before it was all over' seems characteristic of many accounts of 'extraordinary knowing'.

So I can't dismiss the strong possibility that these experiences are indications of an aspect of consciousness or mind that goes beyond the body and brain. In parapsychologist Dean Radin's words, borrowing from quantum theory, psychic or psi phenomena seem to show that consciousness has a 'wavelike' aspect that is somewhat independent of time and space.[28] This 'additional aspect' of consciousness demands a more open approach.

This is all very well, a critic might say, but anecdotes are not proof. One hundred, or one thousand accounts of alleged telepathy, clairvoyance, or precognition are outweighed by one carefully controlled laboratory experiment. One response would be that case studies and collections are significantly more substantial than a random anecdote told over the campfire or in the pub. Here, we've really only dipped our toe into a vast sea of exceptional human experiences. A further response is that modest-scale laboratory experiments investigating alleged psychic abilities have been conducted since psychic research began, and that some of the best results have been very curious indeed.

Chapter 5

Nonlocal Consciousness in the Laboratory?

In spring 2000, I witnessed an experiment that defied conventional science. This was in, of all places, Northampton University. In those days, the psychology department was housed in the Fawsley building on Park Campus at the edge of town. I'd come up by train, taken a bus to Broughton Green road, and walked the rest of the distance into the campus proper. To get to Fawsley you had to walk past parrots in a mesh-fronted cage and, after that, a space-age restaurant. A short set of steps took you down into a little hollow, shaded by pines. Fawsley nestled in this hollow, a single-storey modular building containing classrooms, lecture rooms, and psychology labs.

The building was also the home of the Centre for the Study of Anomalous Psychological Processes (CSAPP).[1] This unit had been set up by Chris Roe. Roe had completed his doctorate at the University of Edinburgh, at the Koestler Parapsychology Unit under Professor Robert Morris. Morris was an American veteran parapsychologist, with a number of decades of experimental work under his belt. Chris Roe was following in Morris's footsteps and was now supervising new doctoral students in psychology and parapsychology.

One of those doctoral students was a friend of mine, Christine Simmonds. Christine was working on one kind of parapsychology experiment, the ganzfeld. The ganzfeld was intended to test two possible modes of psi or psychic ability: telepathy and clairvoyance. The experimental set-up at Fawsley was as follows. A 'sender' would sit in one room at one end of the building, watching a short 'target' film clip in a loop over half an hour. In that time, they'd attempt to 'send' information about that film to the 'receiver'. The 'receiver' sat inside a soundproofed room with a heavy door at the other end of the building. The separation of the rooms and the soundproofing were safeguards against possible 'sensory leakage'. This is where the 'receiver' somehow guesses the target from faint noises or other sensory cues.

Inside the soundproofed room, the receiver relaxed on a comfortable, reclining chair. They had half ping-pong balls taped over their eyes and a soft red light shone in their face. This meant that they'd only be able to see a uniform light pink. The room would be otherwise dark. In the meantime, white noise would be played on headphones into their ears. The sound and the ambient low light was intended to produce an unstructured, uniform stimulation field known as the 'whole field' or *Ganzfeld* in German.

Why was this necessary? Charles Honorton, who'd pioneered the ganzfeld procedure in the 1970s, had noticed that many spontaneous cases of psi seemed to happen in moments of reverie or half-sleep.[2] By contrast, psychic experiences seemed to be inhibited by the general 'noise' of everyday, waking life. Think, for example, of a morning commute to work. You wake up, shower, and listen to the news on the TV while dressing. You talk to your family at breakfast. You walk to work along a noisy, traffic-filled street. All the while your mind is busy with the tasks and challenges of the day. If there is a subtle 'psychic' signal, it will likely be hidden in the background, fogged out by the 'noise' of the working day. The ganzfeld, by contrast, aimed to quieten both the ambient sensory stimulation and relax the mind, allowing any psi signal to 'seep through'.

Christine very kindly allowed me to sit in on one session. During that session, we were sealed in the soundproof room with the participant. After the participant had settled in the reclining chair, he was played a progressive relaxation tape.

Then the session proper began. During the session he was to report any experienced 'mentations', or mental images. Christine was ready with a recorder and notebook. Afterwards these mentations would be examined by a blind judge to see how well they matched the 'target' film clip.

The participant said very little during the session. He reported an impression of 'a nodding donkey or oil derrick'. He also mentioned 'something bright moving over the ground'. Something moving fast. This was pretty much the sum total of what he said. After the session had ended, he was played four film clips on the TV that stood in the corner of the room. Only one of these four clips showed the target. One of the possible targets was a brief clip from a movie I recognised: *2010: Odyssey Two*. This movie, a sequel to the science fiction classic *2001: A Space Odyssey*, features a return visit to Jupiter by a Soviet spacecraft that has a rotating middle section. The spacecraft has an industrial design, and the movement of the middle section is indeed very similar to an oil derrick. In the second part of the clip, the spacecraft uses a technique called aerobraking to slow its orbit around Jupiter. This involves entering Jupiter's atmosphere in a rapid deceleration. At this time, the spacecraft burns like a comet: like, in fact, 'something bright moving over the ground'. What surprised me was that the participant said very little, and that what he said seemed to match up very well with the *2010* clip. He chose the *2010* clip without hesitation.

'That's incredible!' I said, after the participant had left. Christine replied that she'd had more exact hits than that. She also mentioned that often it seemed possible to tell the difference between a 'hit' that occurred by chance and when someone seemed to be picking up information. I should add that the participant's guess was correct: that the 'sender' had indeed been watching the clip from *2010*. None of the other clips had come close to what he'd 'seen' with his mind.

Beating Chance?

One isolated experimental session does not prove very much. It's always possible to argue that the apparent correspondence between the participant's words and the target clip were fortuitous, and that they'd made the correct selection by chance. The solution to this is to run lots of experiments and to use statistics to see whether chance is being beaten. This is a basic technique in science, but how does it work?

Imagine that you're sharing a flat with one other person. Each night one of you has to do the washing up. The chore is assigned by the flip of a coin. If the coin comes down showing heads, it's your turn. If it's tails, your flatmate is doing the washing up. Suppose that over time you notice that you're doing the washing up more than your friend. Not every night, but enough for you to be suspicious. You suspect that the coin is doctored in some way so a flip produces more heads than tails. However, when you confront your flatmate, they shrug and claim it's 'just chance'.

How would you know whether the coin really was doctored? One way would be to flip a coin a number of times, noting down the results. The chances of a coin producing heads per flip is 1 in 2. That's a 50% chance, or more technically, a 0.5 probability. Suppose that you flip the coin ten times. The coin produces six heads and four tails. Can we then conclude that your flatmate is cheating? The short answer is no. Although we've come up with more heads than we'd expect, it's actually not above chance level: the chances of getting six heads out of ten are 20.5%, or a probability value of approximately 0.2.[3]

You'd actually have to come up with at least eight heads on a ten-flip run before you could begin to conclude that there was an extra-chance factor involved, and even then it would be marginal. The probability against chance of flipping eight heads would be about 0.04. This is just below the probability of 1 in 25, or 0.05, which is the threshold that in psychology marks the line between chance results and a possibly genuine statistical effect.

However, a scientist wouldn't be satisfied with a run of ten. They'd want to be as sure as possible that a high number of heads was not a statistical fluke. One answer would be to run lots of tests. Say you performed 100 coin flips and came up with 55 heads and 45 tails. You might think that this would be insignificant, given that last time it took eight out of ten heads to produce an extra-chance result. But you'd be wrong: 55 heads out of 100 is actually just statistically significant, with a probability value of 0.048, within significance. And if you got 60 out of 100 runs, then the probability against chance is 0.01, which is getting on for highly significant. This shows the power of large numbers of tests.

One of the things that parapsychologists are looking for when they experimentally test psi are odds against chance. In the ganzfeld, the receiver is shown four target images or films. A 'hit' marks a successful choice, and a 'miss' is an unsuccessful one. By chance alone, you'd expect them to 'hit', or pick the correct 'target' 1 in 4, or 25% of the time (0.25). So if there was an extra-chance factor involved in the receiver's choice of target, over many runs we'd be looking for an aggregate of 'receivers' picking the target more often than 25% of the time.

Here's the first difficulty of testing for psi. Sometimes experiments seem to get extra-chance results and other times they don't. A number of ganzfeld studies have been conducted at Northampton since the late 1990s. Some of them got positive results; some of them didn't. Unfortunately, the 'hit' that I observed went into one of the studies that did not achieve significance. However, another, more successful study at Northampton reported a hit rate of 35% and was above chance expectation. Overall, too the combined hit rate of all ganzfeld studies at Northampton up to 2024 is significant, at 32.1%.[4] But the fact is that quite a few parapsychology experiments have not produced more than chance results.

This could be for a number of reasons. The first possibility is that the experiments demonstrate that psi does not exist. Any extra-chance results would in that case be flukes, even the experimental runs that did manage to reach experimental significance. Another possibility is that any success was due to some flaw in the experiment. Perhaps the experimenters were doing something, or failed to do something, that meant that receivers were able to guess correctly the correct target. Perhaps, for example, in some experiments they could hear the sound of the video. This is known as sensory leakage. Finally, in the ganzfeld there had been one — admittedly unsubstantiated — accusation of experimental fraud.[5]

Parapsychology has faced such critiques for many decades. When experimental parapsychology began in the 1930s, critics tried to invalidate positive results by questioning the statistical methods that were used and by pointing to possible experimental flaws.[6] But by the 1980s, skeptical critiques had shifted. By then, the focus was on possible problems like selective reporting and multiple analysis. Selective reporting means that only successful experiments get published and failures get discarded, biasing the overall results. Multiple analysis meant that several of the investigators had used different kinds of statistical tests on the parapsychological database, possibly inflating the impression that any effects were significant.[7]

In the 1980s, the main critic of the ganzfeld was the skeptical psychologist Ray Hyman. In a debate with Charles Honorton in the *Journal of Parapsychology*, he pointed to a number of possible flaws in the ganzfeld.[8] However, eventually it was agreed that the positive ganzfeld results were likely not due to sensory leakage, or chance, or selective reporting.[9] Despite this, Hyman thought that there might be a problem with how the target images were selected. He thought that if they were not chosen truly randomly, then this might bias the statistics.[10]

Then Hyman met Charles Honorton at a conference, and they decided to work together on an agreed set of protocols that future experiments would need to follow. In 1986, these protocols were published in a joint communiqué published by the *Journal of Parapsychology*. In this communiqué they wrote: 'We agree that there is an overall significant effect in [the ganzfeld data base] that cannot reasonably be explained by selective reporting or multiple analysis.'[11] The communiqué became the standard for quality in the ganzfeld experiments of the 1990s and early 2000s. Another development was the *autoganzfeld*, where targets were randomly selected and presented by computer.

After the communiqué, the main issue became whether the ganzfeld effect could be successfully replicated. Replication is when several different scientists are able to perform the same experiment and attain comparable results. It's a crucial part of science. One way that the replication issue can be assessed is with a special kind of statistical test called a meta-analysis. A meta-analysis pools different studies of the same experimental technique in order to see whether any significant effect is replicated across studies. This increases what is known as statistical power. So instead of having say 25 or 50 experimental runs, you can see whether an effect replicates over hundreds or even thousands of trials. This means that even small effects, or deviations from chance, can potentially be detected.

In 1994, the *Psychological Bulletin* published a meta-analysis of many ganzfeld studies by Charles Honorton and another psychologist, Daryl Bem.[12] This was a significant coup for parapsychology. The *Psychological Bulletin* is a major scientific journal. The paper carried out a fresh analysis of the studies in the ganzfeld database, which came to 354 sessions in total over all experiments. However, two studies were analysed separately, so they looked at 329 sessions in the meta-analysis.

The 'hit' rate overall was 32.5%, which had significant odds against chance, with a probability of 0.009. For a few years, this was the gold standard of evidence for parapsychology.

Then, in 1999, Julie Milton, with the skeptical psychologist Richard Wiseman, published an update of the ganzfeld meta-analysis in the same journal.[13] They looked at the ganzfeld experiments that had been performed since 1994 and found only chance levels of significance. This analysis caused some consternation in the parapsychology community at the time: I recall one speaker at a conference getting visibly upset at the result. However, this analysis was criticised because they included 'non-standard' ganzfeld experiments in the test, including one where music instead of images had been used.[14] In addition, they left out one highly significant study by Kathy Dalton who'd got a hit rate of 47%.[15]

Lance Storm's 2010 analysis in the *Psychological Bulletin* examined post-1997 studies to 2008.[16] This also excluded Kathy Dalton's study. This new analysis of 30 studies found a hit rate of 32%, which was significant. In 2013, a further meta-analysis by Bryan Williams looking at 59 studies, including 30 from the Milton and Wiseman study and 29 additional ones from Storm's 2010 study, found a hit rate of 30% to 32%. Williams stated that 'the psi ganzfeld effect has indeed been replicated by "a broader range of investigators" under stringent standards'.[17] And the ganzfeld is by no means the only experimental technique in parapsychology to have produced results.

The Remote Viewers

Shortly after 19:00 on 21 December 1988, Pan Am Flight 301, en route from Frankfurt to Detroit, exploded over the Scottish town of Lockerbie.[18] The bomb that had been planted on board killed all 243 passengers and 16 crew. At the time, information on this act of terrorism was sparse. The following is from the files of the CIA Stargate material of operational remote viewing:

Pan Am Flight 103 (Session 7)[19]

Task: *Access and describe cause of the Pan Am crash Flight 103 over Scotland, killing about 273 persons. Trace the cause of the crash to its origin and describe your findings. Describe the person(s) or group(s) involved. And describe the origin of the cause of the crash....*

Summary of information: *The target resembles a 'tent for rent' with a symbol. It is carried in a 'box car/shuttle' (baggage). The object 'has a gunning effect—it wipes out' (explosion?) There are perceptions of a cemetery, a base, (a group of) 'girls' and 'servicemen'. 'This is a plane crash—shameful—terrorism—Flight 301 (Flight 103). The bomb was placed in a communication device by a little, short (even somewhat gentle) person who was aware of the consequences of placing a bomb aboard the plane. This person will be apprehended and brought into custody in England....[This is a summary prepared by the operations officer from notes taken during the session].*

Assessment: *Confirmed.*

The above report is taken from a then classified research programme sponsored by the US government project named Stargate. This was a 23-year programme that intended to assess the use and validity of what was called 'informational psi'.[20] This programme began in the early 1970s at the Stanford Research Institute (SRI). At the time, the main investigators were the physicists Hal Puthoff and Russell Targ. Puthoff and Targ worked with a number of remote viewers who seemed — in a normal, waking state of consciousness — to be able to elicit information from distant events and places.[21]

One of these remote viewers was named Ingo Swann, who was dubbed the 'armchair traveller'. Swann had written to Hal Puthoff in 1972 describing his work with a psychologist, Gertrude Schmeidler, on psychokinesis. Puthoff invited Swann to SRI for a week of experiments. Some of the results seemed interesting, and Swann was invited over to SRI for a longer study. During this time, in between more formal experiments, Swann suggested a 'game'. Puthoff would give Swann a target map coordinates, latitudes and longitudes, and Swann had to state immediately what he perceived. He appeared to be able to do this with uncanny accuracy, so they arranged a further, more formal test. They received some map coordinates sent by a scientist who was challenging their psi experiments. Swan was given these and had to reply immediately. He said:

> My initial response is that it's an island, maybe a mountain sticking up through cloud cover. Terrain seems rocky. Must be small plants growing there. Cloud bank to the west. Very cold. I see some buildings rather mathematically laid out. One of them is orange. There is something like a radar antenna, a round disc [...]. Two white cylindrical tanks, quite large. To the northwest, a small airstrip. Wind is blowing. Must be two or three trucks in front of the building. Behind, is that an outhouse? There's not much there.[22]

The coordinates were for Kerguelen Island in the Indian Ocean. At the time it was a base for a French-Soviet research facility conducting meteorological experiments. Swann also produced a map that was broadly accurate. This was the first of a run of more formal tests of coordinate remote viewing.

SRI did remote viewing experiments until 1990, which were funded by the US government. Later on, from 1990 to 1995, further tests were carried out by Science Application International Corporation (SAIC).[23] These experiments, at least to the experimenters, demonstrated 'the utility of informational psi – the acquisition of non-inferential information from a distant point in space-time'.[24] One reason why they felt they could state this with such confidence is that the remote viewing techniques seemed to provide practical, useful intelligence information on a repeated basis.

The information provided by a remote viewer on Pan Am Flight 103 is only one of a large number of experimental tests of remote viewing. Very often, these would be on classified targets: the information could be at times highly accurate. A remote viewing session on a target place, which the viewer was asked to scan at 2-weekly intervals in 1980, yielded information that was assessed by a CIA analyst as 'highly accurate and of value for operational planning'.[25] At other times, however, there were omissions or mistakes. The remote viewing of a Soviet agent returned several correct details, but 'key questions of interest to intelligence community remain unanswered'.[26]

The Fate of Stargate

At the start of 1995, Congress directed Stargate to be transferred to the CIA.[27] This transfer included a review and technical analysis, which was performed by the American Institutes for Research (AIR). Several experts were called together in a review, including professor of statistics Jessica Utts and skeptic Ray Hyman. The panel had to answer four questions: (1) Did the evidence indicate a statistically significant effect? (2) Can the effects be attributed to a paranormal phenomenon? (3) Is there an adequate understanding of the phenomenon? (4) Does the research provide support for gathering intelligence?[28]

The conclusions were, for question (1) that, yes, an anomaly that was statistically significant had been established, but that (2) the remote viewing phenomenon was not convincingly paranormal, that (3) the nature and source of the anomaly had not been identified, and finally (4), that there was no evidence it had been use for gathering intelligence.[29] These answers might seem to the reader paradoxical and contradictory. The panel was acknowledging that yes, an anomalous effect was found, and that its nature was unknown, but that they could rule out a paranormal effect. And the conclusion that it was of no use to intelligence might also seem to be contradicted by the confirmation by CIA analysts that some of the information was verifiable and confirmed.

It's quite likely that the AIR panel did not have enough information to reach a full conclusion on Stargate. Writing at the time, the programme director of Stargate, Ed May, stated that the scientists evaluating the research were given restricted information.[30] Only a small portion of the operational remote viewing database was examined. Furthermore, the AIR panel did not speak to 'significant program participants' which may also have biased their conclusions. The data restriction precluded 'positive findings. The CIA did not contact or ignored people who possessed critical knowledge of the program, including some end users of the intelligence data.'[31] So it's quite possible that the conclusion that remote viewing was of no value for intelligence was wrong.

Part of the reason for this conclusion was that different panel members drew diametrically opposite conclusions about the data. The statistician Jessica Utts later stated that, in her view, a psychic functioning effect of a small to moderate magnitude had been demonstrated:

> Using the standards applied to any other area of science, it is concluded that psychic functioning has been well established. The statistical results of the studies examined are far beyond what is expected by chance. Arguments that these results could be due to methodological flaws in the experiments are soundly refuted. Effects of similar magnitude have been replicated at a number of laboratories across the world. Such consistency cannot be readily explained by claims of flaws or fraud.[32]

Ray Hyman disagreed from this conclusion, offering the following thoughts in a later paper review:

> My report argues that Professor Utts' conclusion is premature, to say the least. The reports of the SAIC results and those of other parapsychological experiments may be illusory. Many important inconsistencies are emphasised. Even if the observed effect could be independently investigated, much more theoretical and empirical investigation would be needed before one could legitimately claim the existence of paranormal functioning.[33]

However, elsewhere he explicitly stated that Utts and he agreed 'on many points', that the revised experiments were free of 'methodological weaknesses' and 'obvious and better known flaws'. He also conceded that 'the effect sizes reported [...] are too large and consistent to be dismissed as statistical flukes'.[34] This was not the end of the critical backlash. A later piece by the critics Julie Milton and Richard Wiseman suggested that some crucial data was missing from the final reports on the first SAIC experiment, which in their view meant that alternative, non-psi explanations could not be ruled out.[35] In 1995, Stargate was closed down, which might superficially seem to confirm the problems with the data.

However, this would not be an accurate conclusion. Marwaha states that the CIA was suffering from scandal when the agency gained Stargate.[36] The CIA was therefore under political pressure and was downsizing in a post-Cold War context. Stargate fell victim to these budget cuts.

A 2023 systematic review and meta-analysis of remote viewing from 1974 to 2022 found a 'strong average effect size' over 36 studies. The conclusion was that 'After more than 50 years of investigation into extrasensory perception, remote viewing experimental protocols appear to be the most efficient for both experimental and practical applications.'[37]

So Has Psychic Functioning Been Demonstrated?

So has psychic functioning been demonstrated in the laboratory or not? I think we can say with some confidence that despite the important objections of the critics, there exists in the parapsychological database persistent statistical anomalies that do not seem to be the result of any obvious experimental flaws or statistical errors. This might sound quite dry, but it implies quite a lot. For me, it tips the scales significantly in favour of psychic functioning.

This conclusion has recently been backed up by a couple of new meta-analyses. One by Patrizio Tressoldi used a special kind of Bayesian meta-analysis to look at seven different types of lab experiment for psi.[38] He found significant statistical effects in all seven databases. In 2018, the American Psychological Association published a paper by Ertzel Cardeña that summarised the experimental evidence for psi with another meta-analysis. He concluded that overall the results 'provide cumulative [...] evidence for psi'.[39]

What about the objections of the critics? I think the main problem does not lie with the reliability or otherwise of the positive results, experiments, or statistics. I think the main problem is that psychic functioning seems to be at odds with the assumptions of scientific materialism. The prominent critic of parapsychology, Richard Wiseman, more or less admitted as much when he stated that 'by the standards of any other area of science [...] remote viewing is proven'.[40] This is a significant concession because it tells us that a major reason for resistance to the idea of psychic functioning is that it's considered vanishingly unlikely or even impossible. Chris Roe noted that there remained a tendency for people to evaluate the quality of research 'on the basis of their prior beliefs'.[41]

For me, the remote viewing and ganzfeld databases contain some of the best evidence that psychic functioning is real. Although this is an admittedly subjective impression, I find it hard to believe that the direct 'hit' that I observed in Northampton was purely fortuitous. The information the receiver was receiving seemed too specific. Of course, one subjective impression on its own shouldn't persuade anyone. But it fits in within a broader sweep of evidence that seems to me strongly suggestive of some kind of anomalous process of information transfer that defies time and space.

At the same time, I share the concerns of the critics over the replication problem. If this anomalous form of information transfer is real, then why can't it be more widely replicated? If psi is a fish, then our nets currently seem full of holes. It seems to me that the best way forward is to get a better handle on the so-called 'experimenter effects'. According to some researchers, if psi really does exist, then this means that human consciousness can interact with its environment beyond the usual space-time boundaries, and we might expect significant experimenter effects. This is because, if psi were real, the experimenter would be an integral part of the experiment, and their consciousness would also be expected significantly to influence the outcome.[42]

But maybe we don't need to invoke psi to account for experimenter effects. It turns out that other kinds of experiments in psychology and medicine can also be influenced by human factors like interpersonal interactions and personality. In the 1960s, the psychologists Robert Rosenthal and Lenore Jacobson showed that teachers' expectations seemed to have an effect on how well their students did in the classroom.[43] This 'Rosenthal effect' is probably familiar to many of us from school days, when a teacher's attitude had a significant influence on how we'd perform.

In this chapter, I do not hope to have persuaded the reader of the reality of psychic functioning. However, I do hope that I've shown that there's more than enough to justify the existence and continuation of experimental parapsychology. Parapsychologists, in steadily building a robust database over many decades with often meagre resources and in defiance of strong criticism, at the very least deserve a fair hearing. But personally I find it difficult to believe that these results are 100% the result of error. Psi in some form seems real.

This is a tricky conclusion to reach in a culture where hard-line skepticism is the official default. Parapsychology remains siloed off from 'respectable' topics in psychology and neuroscience. Chris Roe argues that this partition is unjustified, urging everybody in the debate to 'set aside [...] prior beliefs and prejudices and to make a judgement on a claim based solely on the weight and quality of the research evidence'.[44] This should remain the ideal in any public discussion of the evidence for psychic functioning. Unfortunately this is especially tricky to do because serious work in parapsychology remains opposed by vocal networks of vociferous skeptics who police the boundaries of science. And it's to the arguments of skeptics that we'll turn next.

Chapter 6

Skepticism, Rational and Irrational

The early 1970s:

Susan Blackmore arrived at St Hilda's College, Oxford to complete an undergraduate degree in psychology and physiology. She already harboured a new obsession. The previous summer she'd read Rosalind Heywood's *The Sixth Sense*, an introduction to parapsychology. A causal interest quickly turned to fascination. She was, in her words, 'hooked':

> Parapsychology has everything a hook needs. It is mysterious and alluring. It has enough "scientific" evidence to provide bait, while at the same time is rejected by most orthodox scientists, the inspiration for a crusading spirit to shout "I'll show them."[1]

Her experiences are candidly recounted in her book *In Search of the Light: The Adventures of a Parapsychologist*. When she got to Oxford, she helped revive the moribund Oxford Psychical Research Society. Under the influence of marijuana, she also experienced a very powerful out-of-body experience (OBE). In this experience, she apparently left her body, viewing it from above. She floated out of the building and journeyed through a luminous, vibrant otherworld.[2] The experience seemed utterly real. By contrast, she found the everyday psychology and physiology of her degree dry and uninspiring. The powerful OBE deepened her obsession, acting as a trigger to put most of her energies into finding psi. She did this despite her tutor urging her to quit her obsession and despite the *Bulletin of the British Psychological Society* publishing a humiliating review of one of her lectures.[3]

She was also unable to find any evidence for psi that satisfied her. Her first experiment on psi and memory was a total disaster. In subsequent experiments, she also seemed to find nothing. This pattern continued through her masters and doctoral programmes at the University of Surrey. Repeatedly, Blackmore sought evidence for the paranormal and, repeatedly, came away disappointed.

Additionally, she'd become critical of other people's parapsychology experiments. She believed that Ernesto Spinelli's experiments with psi in children had significant flaws. She thought that errors in experimental design might explain the positive results that Spinelli had reported. Later on, she visited Carl Sargent's lab in Cambridge. Sargent had been getting apparently spectacular results with his ganzfeld experiments. Blackmore went to try to understand why Sargent was getting positive results and why she wasn't. She eventually concluded that he was cheating.

This failure to find psi seemed to take its toll on her physical health. She lost weight and was eventually diagnosed with a hyperactive thyroid. This period, around 1980, was also one of psychological turmoil. Slowly, by degrees, she was abandoning psi, calling it a 'failed hypothesis'. She also re-assessed her earlier out-of-body experience. She came to believe that it could be explained entirely in terms of brain function. For example, in the initial experience, she noted a difference between her perception of the roof of the building she had seen in her OBE and the roof as it was in real life.[4] This seemed to imply that the OBE, although very vivid, was basically hallucinatory.

This psychological shift was very tough for her. In Blackmore's writings, she's very honest about the 'inner battles' that accompanied her psychological transformation from a passionate believer to a skeptic. Towards the end of *Adventures*, she writes about going down onto a Devon beach and mulling over her personal journey:

> Was I really the same person who had held out so much hope for psi? [...] Would I ever know whether there was psi or not? And all those claimed successes that I had met with so often, were they really due to psi, or chance, or bad experiments, or what? I didn't know. I hated not knowing, but I really didn't know [...]. It was obvious, wasn't it? Even the warring factions inside me seemed to agree. *I didn't know.*[5]

In a later article, Blackmore explained why she still considers herself a skeptic. She believes that the answers to many unusual experiences can be found in conventional psychology and neuroscience. She also rejects claims that skeptics are dogmatic:

> Psychics and believers tell me that I don't have an open mind. "Do you know what it means to have an open mind?" I want to shout sometimes. "It means being able to change your mind when the evidence shows you are wrong." That is what science is about.[6]

This is, superficially, a strong point. One primary aim of science is to test our theories of the world against evidence. You set up an experiment, perform it, and record the results. If your results confirm what the theory predicts, then you have evidence. If they don't, then you don't. In theory, it's simple. But in practice, finding the truth, or an approximation of the truth, can be a long and drawn out process. There are many pitfalls, and any conclusions drawn must often be tentative, and with new data can even be reversed.

Consider Blackmore's own experiences. She claimed to have found evidence of fraud in Carl Sargent's experiments. Eventually, these accusations were published in the *Journal of the Society for Psychical Research*.[7] Sargent wrote a reply denying the allegations, and I suggest that anyone interested in the controversy read the original documents for themselves.[8] In any case, Blackmore's allegations were never proved. Chris Roe has recently found problems with Blackmore's analysis. He writes that Blackmore's concerns are 'technical and quite complex' and possibly wouldn't have worked the way that Blackmore claims. He also suggests that many of the complexities in the debate have subsequently been substituted for brief, definite conclusions. The definite conclusions 'are not justified by the material' that Roe reviewed.[9] So the fraud claims are questionable.

Blackmore's other claim, that she did not find psi in her experiments, has also been questioned. The parapsychologist Rick Berger reviewed Blackmore's experiments, including unpublished experimental data. He claimed to find discrepancies between the raw and published results, and suggested that 'flaws' had been engineered to dismiss significant results, but other flaws in nonsignificant experiments were ignored. He suggested that Blackmore's change of viewpoint had occurred over less time than suggested in her book, probably over 2 years, on the basis of experiments conducted between October 1976 and December 1978. Also, 30% of the experiments were successful, contradicting her claims of uniformly negative results.[10]

So things are more nuanced than they might at first seem.

Skeptics: Two Strategies

Skeptics have used two strategies to cast doubt on psychical research. The first strategy might be called *informed critique*. Here, skeptics offer criticisms of experiments and try to find conventional explanations for apparently 'paranormal' events. These sorts of criticisms are often intended to be constructive. This first strategy is for the most part that followed by Susan Blackmore.

Constructive criticism is essential for science. Any scientist worth their salt will want to discover experimental flaws and eliminate them. If an experiment is flawed, then any positive results might be due to those flaws. Similarly, if apparent ghost sightings or telepathic experiences really have conventional explanations, then it's surely better to know it. Having a conventional explanation for unusual experiences is also a gain in knowledge.[11] For a scientist, all knowledge is valuable. So this first kind of criticism is very important and also in many ways a standard part of how science works.

We've already seen that this first kind of criticism has been very useful for parapsychology. The debates over the ganzfeld between Ray Hyman and Charles Honorton led to a joint communiqué that was used to resolve doubts about the methods used in earlier ganzfeld experiments. The communiqué helped significantly to improve the autoganzfeld, the computerised version of the experiments. Similarly critiques of the remote viewing experiments also helped to improve the methods used in those experiments. So parapsychology has in the long run benefited significantly from criticism.

However, skeptics now commonly employ a second, more destructive strategy. This is where parapsychology is dismissed or ridiculed as 'woo-woo' or worse. Critics have tried quite a few strategies to show that parapsychology is not really science.[12] They have claimed that it is magic and so the opposite of science. They've stated that the claims made by parapsychologists violate the laws of physics, so they must be impossible. They've pointed to parapsychology's association with the New Age movement. They've also claimed that parapsychologists are on an 'irrational' quest for the soul.[13] They've used mockery and ridicule. They've attempted to censor online talks and strip researchers of their qualifications.[14] And they've pursued high-profile media campaigns that attempt to discredit any and all of the claims of parapsychology.

This second type of criticism has been called *boundary work* by the historian Thomas Gieryn.[15] 'Boundary work' is not actually necessarily destructive. It's when you decide, fairly arbitrarily, what is 'science' and what is 'pseudoscience'. If you decide that a topic is 'scientific', then you can advocate for its study. This is essentially what parapsychologists do, and also in a sense what this book is doing. When I claim that psychic experiences are a worthy topic for scientific investigation, I'm engaging in boundary work. I'm claiming that studying psychic experience can in principle be as 'scientific' as studying atoms, brains, or planetary orbits.

Modern skeptical advocates push in the opposite direction. The main aim is to prove that parapsychology is not 'science' at all but 'pseudoscience' and so beyond the pale. This is harder than it looks. It raises questions about how 'real' science might be distinguished from so-called pseudoscience. There have been quite a few attempts to find systematic differences between real 'science' and 'pathological' or 'voodoo' science. None have been totally successful.[16] This is because mainstream science often has elements of what have been called pseudoscience and vice versa. Secondly, on occasion, fringe topics have eventually been accepted as science. Examples include the extraterrestrial origin of meteorites and continental drift.[17]

Because of this difficulty, skeptics often adopt a restrictive set of assumptions about how the world works. Essentially they end up defending the current mainstream consensus in science. The sociologist Brian Martin put it this way:

> The Skeptics movement is held together by a common set of assumptions about the world and their place in it. Skeptics assume they have access to the truth (even if it is hypothetically provisional), which is almost always based in scientific and medical orthodoxy, and that those who have different beliefs about core issues are wrong and need to be educated or countered. These assumptions come to the fore in Skeptics [*sic*] explanations for

> heterodox beliefs, which they attribute to psychological shortcomings [...].[18]

This second skeptical strategy of aggressive boundary work dominates in the social media era. This is partly because of the current cultural context. We live in a globalised society that has become, politically speaking, highly polarised. There's evidence that the learning algorithms of social media have helped worsen this polarisation by funnelling people into different bubbles of mutual suspicion. This change has, along with everything else, also affected public disputes over parapsychology. Having said this, the basic 'boundary work' strategy of organised ridicule and aggressive debunking predates social media, having its origins in the early 1970s.

Modern Skepticism's Origin Story: The Case of Uri Geller

In the early 1970s, a young psychic named Uri Geller began to attract significant media attention.[19] Geller was born in Israel in 1946 and exhibited apparent psychic abilities from an early age. He seemed to be able to advance the hands of clocks without touching them and also claimed to be able to reproduce drawings from a distance. His most famous feat was the apparent ability to bend metal with the power of his mind. For the most part, his public performances focused on cutlery like spoons or forks. This inspired a label for the subsequently inspired craze: 'spoon bending'.

Geller attracted both publicity and controversy from the beginning of his public career. He became a full-time entertainer in early 1970. By October of that year, *Haolam Hazeh* magazine put a photo of Geller on its cover with the announcement that 'all of Israel's magicians have assembled for a witch-hunt' against 'this telepathic impostor'.[20] Despite the 'witch-hunt', Geller had attracted the attention of the researcher, Andrija

Puharich, who from 1971 tested him numerous times and found his abilities convincing.[21] Puharich alerted other interested scientists to Geller and subsequently several tested him. At the same time, Geller was getting a lot of attention from media in Europe and America, appearing in many newspaper articles and on television. He was in a critical article in *Time*. He also appeared on many US talk shows, like *The Merv Griffin Show* and *The Tonight Show*.[22]

The peak of the scientific attention given to Geller was the publication of an article in the highly prestigious scientific journal *Nature* in October 1974.[23] The article was published by Harold Puthoff and Russell Targ, the same researchers who'd initiated the remote viewing programme at Stanford University. Puthoff and Targ had also been alerted to Geller's apparent abilities by Puharich and, despite some skepticism, had agreed to test him at SRI. Targ and Puthoff later reported that when Geller came and spent about 6 weeks at SRI, they 'saw many wild and wonderful things', but despite shooting 30,000 ft of 16 mm film, they did not ever 'photograph what we had set out to observe'.[24] Despite this, they were also able to test Geller's remote viewing abilities, and it was this that was the main focus of the *Nature* article.

According to the historian of science Jacob Older Green, this article marked the peak of mainstream scientific interest in parapsychology in the late 1960s and early 1970s, as well as the beginning of the modern skeptical movement.[25] It was also the moment that skepticism moved from a strategy of significant, constructive criticism to a much stronger strategy of aggressive rejection or 'boundary work'.

From the dawn of experimental parapsychology in the1930s to about the 1970s, critics had tended to see parapsychology as a flawed but legitimate scientific exercise. Many of the critiques in this earlier time pointed to possible flaws in the experiments or statistics. At the same time, especially in the

1960s, parapsychology received some official funding. This included Targ and Puthoff's work at Stanford.

This situation changed after the publication of the *Nature* paper. In a letter to the *New Scientist* the medical researcher and science journalist Donald Gould suggested that Geller's alleged powers had reduced 'normally hard-headed observers of the human scene to the status of awe-struck peasants'.[26] Joseph Hanlon suggested that the researcher's 'desire to believe' might have 'clouded their discrimination'.[27] Meanwhile, Maurice Wilkins, a Nobel Prize winner, witnessed James 'The Amazing Randi' reproduce many of Geller's feats at King's College, London, and subsequently drafted a joint letter with other scientists, insisting that magicians should be included in any subsequent investigation. The signatories included the editor and deputy editor of *Nature*.[28]

It's important to understand the wider context of this dispute. The 1960s counterculture had happened, and young people were questioning the establishment in many ways. In America, there was the civil rights movement and protests against the Vietnam war. But the questioning went deeper than that. Many young people challenged the basic institutions of industrial consumer society. A key feature of the counterculture was the rejection of scientism and industrial technologies and the embracing of indigenous traditions and eastern mysticism. Psychedelic drugs, particularly LSD, became popular as a method of expanding consciousness. There was also an occult revival, which many scientists perceived as a threat.

The concern was that young people were rejecting 'science' and 'reason' and embracing magic and superstition. The astronomer and science populariser Carl Sagan later expressed this fear in his book *The Demon-Haunted World: Science as a Candle in the Dark*. He foresaw a time when 'clutching our crystals and nervously consulting our horoscopes, our critical faculties in decline, unable to distinguish between what feels good and what's true, we slide, almost without noticing, back into superstition and darkness'.[29] Although written in the 1990s, in this passage Sagan was expressing the basic fears of many scientists since the 1970s.

However, it's possible that those fears were displaced. Green suggests that in fact the public didn't reject science per se, but that this interest often went hand in hand with a rise of interest in occult topics. But the perception of threat remained and this plus the arrival of Geller was one of the spurs to the creation of the modern skeptic movement. The magician James 'The Amazing Randi', who'd so impressed Maurice Wilkins, was one key founder of this movement.

Randi was a colourful and theatrical character. Born Randall James Hamilton Zwinge on 7 August 1928, in Toronto, Canada, he once played an executioner on an Alice Cooper tour, pretending to behead Cooper with a guillotine.[30] He was also an avowed atheist, stating that 'a belief in a god is one of the most damaging things that infests humanity at this particular moment in history'.[31] His feelings about the paranormal were similarly hard-line, dismissing any and all claims as nonsense. From the mid 1970s onwards, he became Geller's nemesis. He claimed to be able to reproduce all of Geller's performances by conjuring.

Randi was also one of the co-founders of the Committee for the Scientific Investigation of Claims of the Paranormal (CSICOP). CSICOP was the direct descendent of an earlier, informal group called Resources for the Scientific Evaluation of the Paranormal, or RSEP.[32] RSEP consisted of magicians, including Martin Gardner, Ray Hyman, James Randi, and Marcello Truzzi. Several of the members were also academics; we've already encountered Ray Hyman as a prominent critic of the ganzfeld and remote viewing experiments. Together with the philosopher Paul Kurtz and other affiliates like Carl Sagan, this group formally founded CSICOP in 1976. The aim was to provide a critical perspective on the occult revival and subjects like Bigfoot, Atlantis, the Bermuda Triangle, and UFOs.[33] The organisation still exists today, although it was renamed the Committee for Skeptical Investigations (CSI) in 2006. Its journal is the *Skeptical Inquirer*.

From early on, there were differences between the members on how to approach these topics. Marcello Truzzi, who was a sociologist as well as a magician, was initially the editor of CSICOP's magazine. He favoured dialogue and debate with advocates of parapsychology and other contentious topics. However, Paul Kurtz, Randi, and others favoured science advocacy. This meant taking a 'boundary work' perspective and actively debunking topics like parapsychology. This stance was reinforced by the hard-line presence of Randi, and the writer Mitch Horowitz has suggested that Randi's approach was in effect adopted wholesale by the modern skeptical movement.[34] So denouncement and debunking were in, and dialogue and debate were out. This approach didn't suit everyone: Marcello Truzzi resigned from CSICOP in August 1977. But it's fair to say that the hard-line, debunking stance that CSICOP adopted from the early days still shapes skepticism to this day.

Twenty-First-Century Skepticism

The most visible face of skepticism today is on Wikipedia. If you Google search 'parapsychology', the top return is the Wikipedia page. In the first paragraph, Wikipedia states that:

> Criticized as being a pseudoscience, the majority of mainstream scientists reject it. Parapsychology has also been criticised by mainstream critics for claims by many of its practitioners that their studies are plausible despite a lack of convincing evidence after more than a century of research for the existence of any psychic phenomena.[35]

These are standard skeptical criticisms about parapsychology, and a little later I'll address each one. But anyone who was not aware of the background of the debates may well take these claims at face value. They might, in other words, accept that parapsychology lies beyond the boundary of acceptable science. But they wouldn't do this because they've looked at the evidence and found it wanting. They'd tend to do it if they accepted that Wikipedia was an authoritative and disinterested source of factual information.

But things are not quite what they seem. Wikipedia can be edited by anyone, and this has made it vulnerable to being hijacked by special interest groups. For example, pages on contentious topics like parapsychology and alternative medicine have been targeted by a group called 'Guerrilla Skepticism on Wikipedia'. These 'Guerrilla Skeptics' are alleged to police topics like parapsychology, which includes the biographies of researchers. The *Skeptical Inquirer* reports that 'The Guerrilla Skepticism on Wikipedia (GSoW) project was started in May 2010 as an effort to unite editors to become more skilled at adding skeptical content to the fifth most popular Internet site in the world.'[36] There have even been claims that hundreds of Wikipedia pages are monitored and aggressively edited to conform to a hard-line skeptical narrative.

This approach mirrors the 'scorched-earth' tactics of James Randi, which isn't surprising. Jerry Andrus who helped the Guerrilla Skeptics set up the first page for their world Wikipedia project has been described as a 'quintessential skeptic and great friend of Ray Hyman and James Randi'.[37] Susan Gerbic, a studio photographer who founded the project, stated in 2013 that 'It is, in my opinion, probably the farthest reaching and measurable world-wide skeptical project that exists today.' And that with training, anyone of a skeptical persuasion can participate 'from home in your pajamas'.[38] This project began as a way of policing psychics' Wikipedia pages as a way of reducing their credibility, because the skeptical movement saw that having a Wikipedia page was 'a notch-up in credibility for these people'.[39] Unfortunately, this project has ended up failing to discriminate between media psychics and serious scientific researchers.

When the Guerrilla Skeptics project drew criticism, Gerbic and others denied that there was any specific campaign against parapsychology researchers. In a blog post, Gerbic stated that their edits were always transparent and clearly labelled.[40] Specifically, she poured scorn on claims that two figures publicly associated with parapsychology, Rupert Sheldrake and Deepak Chopra, had been targeted by the group. She dismissed these sorts of claims as a 'conspiracy theory'. This particular dismissal might be justified, but it doesn't eliminate the problem of persistent bias on Wikipedia against parapsychological research. Brian Martin has noted several potential sources of bias on Wikipedia. Techniques include 'deleting positive material, adding negative material, using a one-sided selection of sources, and exaggerating the significance of particular topics'.[41]

No doubt skeptically-inclined Wikipedia editors, organised or not, feel justified in doing what they do. But the effect is to 'police orthodoxy', as Brian Martin put it.[42] In their eyes, they're defending science and reason against irrational forces. And to some extent, this attitude is understandable. We live in an age of fake news, conspiracy theories, rumour-mongering, AI 'deepfakes', and information warfare. False claims abound on social media and the World Wide Web. Information has become weaponised, as different interest groups vie for credibility. There has been a resurgence of blatantly false theories, like the flat earth. In this context, 'science and reason' can indeed seem besieged.

Withering Skepticism?

Still, one question remains: is taking a 'scorched-earth' approach the best way to deal with the controversies over parapsychology? Recently, some researchers in the field have been pushing back and offering robust critiques of the current state of skepticism. Chris Roe, now a professor of psychology in the UK, has made some strong critiques of the state of twenty-first-century skepticism in several talks and lectures.

Roe has reviewed several recent, representative skeptical works and found the same overworked arguments in most of them. What skeptics often do is employ what Roe calls identical 'rhetorical ploys' in piece after piece. We've already met one example in the Wikipedia article: claims that parapsychology lacks 'convincing evidence after more than a century of research'. This claim, what Roe has called a 'sweeping generalisation', has been repeated over and over in the skeptical literature. It is demonstrably false, as I hope that this book has already shown. There exists a large database of experimental results with persistent anomalies that skeptics themselves sometimes admit do not seem to be explicable in terms of systematic experimental or statistical errors.

Roe can point to specific examples showing that skeptics are often ignorant of the basic literature. For example, the claim by Michael Shermer that 'Under controlled conditions, remote viewers have never succeeded in finding a hidden target with greater accuracy than random guessing.'[43] A cursory look at the remote viewing database shows this to be completely false. If Shermer's statement were true, then Jessica Utts would never have been able to claim that 'The statistical results of the studies examined are far beyond what is expected by chance,' in her review of the Stargate data. This conclusion, remember, was not in itself disputed by Ray Hyman. As Roe states, 'the idea that Shermer's conclusions are based on a thorough review of evidence is unpersuasive'. He goes on to suggest that 'This type of shallow scholarship cannot be allowed to pass unchallenged.'[44]

There exist more extreme examples of what Roe calls 'eschewing the data'. In 2018, Etzel Cardeña of Lund University published an article on parapsychology in the *American Psychologist*, the prestigious journal of the American Psychological Association. The Association is the leading professional and scientific organisation for psychologists in the USA. This article presented the accumulated results of many decades of parapsychological experiments. Perhaps predictably, Cardeña's findings led to significant debate within the APA, and in addition, the skeptics were quick to respond.

Arthur Reber and James Alcock wrote an article responding to Cardeña for the *American Psychologist*. Reber and Alcock pointed to the 'existential impossibility of psi phenomena' and again reported the claim that there had been 'nearly 150 years of efforts during which there has been, literally, no progress'. They also, unbelievably, refused to consider the data presented by Cardeña at all. Their justification was that the 'Claims made by parapsychologists cannot be true,' and that 'the data have no existential value' and so must be due to errors.[45] This seems a very strong, even fantastic, claim to me. Reber and Alcock are saying that our knowledge of the natural world is complete enough to rule out genuinely novel phenomena. This surely displays an overconfidence in current human knowledge.

In fairness, not all skeptics reject the evidence in this way. The psychologists Chris French and Anna Stone claimed that the 'assumption' that 'paranormal forces do not exist' would be revised if parapsychologists or other researchers 'ever manage to produce compelling evidence that paranormal forces do exist'.[46] In some ways, this is a reasonable position. However, it's clear from the context of the comment that neither French nor Stone consider the current evidence base of parapsychology even remotely convincing. It's unclear what kind of evidence might make them change their minds.

The Trouble with Skepticism

Post-1970s skepticism can boast some success. Skeptics have exposed fraudulent mediums and quack medicine and provided platforms for a critical discussion of misinformation and latterly 'fake news'. The pages of the main skeptics' magazine, the *Skeptical Inquirer*, and the UK and US *Skeptic* have hosted quite a number of convincing critical or debunking articles on a wide range of topics. Skeptics' organisations no doubt provided a rallying point and a community for those who have become disillusioned with religion or with New Age philosophies. And at times, skeptics have worked together with psi advocates in ways that have proved fruitful. So to this extent, organised skepticism has been a success.

However, I think that it's also reasonable to point to some significant failures. One of the consequences of adopting aggressive, strident messaging is the danger that you will alienate a portion of your intended audience. This includes people who have had apparently psychic experiences. As Nancy Zingrone, another critic of modern skepticism, has put it:

> Superficial dismissals do not serve the needs of science, nor do they serve the needs of experiencers. Perhaps some of the wilder forms of public belief in the paranormal arise from weak [skeptical] explanations [...]. It does not take a psychical researcher to know that spontaneous experiences and the people who have them are far more complex than the average critic or TV-debunker appears to believe. The experiencers, their friends and families know this too.[47]

Aggressive skepticism reinforces social taboos around psychic experiences. This makes them harder to report or discuss publicly. Experiencers will know that if they disclose their experience, they could be subjected to ridicule or worse. So a policy of blanket dismissal can get in the way of better understanding the experiences that people have and drive witnesses to silence.

There exist more fundamental problems with the aggressive advocacy of skepticism. Skeptics tend to be so focused on the defence of 'science and reason' that they do not really ask why people might be disillusioned with science and technology. As the sociologist Allan Kellehear put it, 'Skeptics have preferred to see this critical move away from scientific rationalism as a weakness in human nature rather than as a reaction to cracks in the modern scientific dream.'[48] There are a number of rather obvious reasons for this public disillusionment that have very little to do with 'irrationality' and everything to do with the increasingly evident limitations of technological science.

For the moment, I'll park the broader questions concerning 'cracks in the modern scientific dream'. This is in order to focus on one area where mainstream neuroscience has recently been confronted with a panoply of extraordinary and transcendent experiences. The psychedelic revival is now big business. Many laboratories are now investigating the therapeutic effects of drugs that can dramatically shift the consciousness of the user. In doing so, this is forcing a confrontation with sometimes extreme mystical experiences, that for the user can challenge the heart of scientific materialism.

Chapter 7

Touching the Infinite?

6 February 1971:

The Apollo command module, christened the *Kitty Hawk*, was on a return trajectory from the moon.[1] Aboard were three astronauts: Alan B. Shepherd Jr., Stuart A. Roosa, and Edgar D. Mitchell. In the previous days, Shepherd and Mitchell had been climbing in the Frau Mauro region of the moon. Afterwards, tired but jubilant, the two astronauts had returned to the command module via the ascent stage of the lunar module. They had fired *Kitty Hawk*'s main engine, initiating an Earth return trajectory.

On 6 February, Ed Mitchell found himself watching the Earth. The jewelled, cloud-covered, blue and ochre disk floated in the darkness, basking in the light of the sun. As Mitchell watched, he experienced a profound shift in his consciousness. He later described it in the following terms:

> I had completed my major task for going to the moon and was on my way home and was observing the heavens and the Earth from this distance, observing the passing of the heavens. As we were rotated, I saw the Earth, the sun, the moon, and a 360° panorama of the heavens.
>
> The magnificence of all of this was this trigger in my visioning. In the ancient Sanskrit, it's called Samadhi. It means that you see things with your senses the way they are — you experience them viscerally and internally as a unity and a oneness accompanied by ecstasy.[2]

This experience marked a personal turning point for Mitchell. Afterwards, he left NASA and in 1973 went on to found the Institute of Noetic Sciences (IONS). The major aim of IONS was to investigate the mysteries of 'inner space' and consciousness.[3]

As Mitchell remarked, this sort of experience is not new. The *Yoga-Sutra of Patanjali,* which was written in India almost 2000 years ago, also discussed *Samadhi.* It is described as the end-point of a series of spiritual practices that involve the withdrawal of the senses from everyday life. According to the Sutra, one moves through a successive series of states of intense concentration, which leads to an uncovering of the basic ground of reality. These stages are *Dharana* (concentration), *Dhyana* (absorption), and finally *Samadhi* (integration).[4] The goal is a transformation of being, facilitated by a mystical understanding of the true nature of things.

Altered States of Being

Human beings can significantly shift their state of consciousness. This is an everyday occurrence. Human consciousness moves through distinct states over the course of 24 hours, between waking and sleep. Dreaming itself is a very different state of being than ordinary, waking consciousness. In a dream, you can do things that might seem ordinarily impossible, like fly or levitate. When you dream, you in effect enter 'another world' with different rules from waking, everyday reality. This world can at times seem as vivid or 'real' as waking life. The environment is more pliable, and you might encounter people who are long dead, other beings, or even monsters. But dreaming is only one variety of what are termed altered states of consciousness (ASC).

The term 'ASC' was proposed in the 1960s by Charles Tart, who was studying not just waking consciousness but also those states induced by trance or hypnosis and by psychotropic and psychedelic drugs like marijuana, LSD, DMT, and psilocybin.[5] Psychedelics in particular seem to induce extreme altered states, where people can experience a radically changed environment, mystical states, and even otherworldly beings.[6] These experiences often seem loaded with significance, revealing aspects of reality that are normally invisible in everyday states of consciousness.

Psychedelic states seem in some ways related to the kinds of experiences that happen in dreams. The difference is that they seem especially vivid and 'real'. For example, the psychedelics researcher David Luke had a startling experience on a drug called DMT. DMT, or N,N-Dimethyltryptamine, is a powerful psychedelic drug.[7] It is a component of the ayahuasca brew ingested by Amazonian tribes like the Asháninca people, whom Jeremy Narby studied.[8] It's also similar to chemicals in the brain that are related to sleep and dreaming. On one particular 'trip' in India, Luke reported the following:

> I travelled to a secluded beach on the banks of the River Ganges. I prepared myself with an improvised ritual, hoping to gird against whatever lay beyond, and I inhaled a pipe-full of vapours from the foul plastic-tasting resin. Sucked into the space between the pipe and my brain, I found myself breaking through the veil like a gatecrasher into a party of swirling, smiling eyeballs all attached to snake bodies, which were as startled to see me as I was to be there. The whole ordered assortment of eyes and snakes acted as one being. In the brief moment before it reacted to my arrival, I managed to catch a glimpse over what might loosely be described as 'the shoulder' of this strange entity and instantly realised that I had seen something I should not have—a brief glance at the truly forbidden.[9]

This account captures the intense and 'hyper-real' character of psychedelic experiences. But drugs are not the only means of significantly altering one's state of being. Spiritual practices like those described in the *Yoga-Sutra* are an alternative means. There are also various kinds of trance state, induced by drumming, chanting, semi-starvation, or other means. Traditional cultures have used these methods for many thousands of years. In these cultures, they are often seen as a means to communicate with spirits or gods.

All in the Brain?

How are these states of consciousness related to sleep and dreaming? Prehistorian David Lewis-Williams, influenced by cognitive psychologist Colin Martindale and others, has suggested that there are not one but two pathways to an altered state, beginning with waking consciousness.[10] One is the ordinary sleep-wake cycle. At some point in the cycle, mostly at night, we drift into hypnogogia, a state between waking and sleeping. In hypnogogia, we're more prone to hallucinations. We might see bright patterns and shapes flit across our visual field in the darkness. These can be pinpoints of flashing light, spirals or net-like patterns. These are called 'form constants'. We might also perceive more vivid hallucinations, such as eyes, faces, buildings, or animals. Eventually, in the first pathway, we slip into deep sleep and dreaming states.

The second pathway is that of trance. Lewis-Williams calls this the 'intensified trajectory'.[11] Traditionally, this has been induced by the various methods described above. Martindale divided trance states into three distinct stages. Stage 1 is similar in many ways to hypnogogia. You see flashes, spirals, zigzags, net-like patterns, and other 'form constants'. David-Williams called these 'entoptic' because they are formed by processes within the eyes. In stage 2, participants try to make sense of the patterns that they see in terms of everyday objects. This can sometimes 'push' the abstract patterns to be perceived as hallucinatory objects. For example, in one especially intense hypnogogic state, I remember grid-like patterns resolving into the front of a house.

The third stage involves a marked shift in consciousness. Lewis-Williams suggested that this was often marked by the arrival of a vortex-like tunnel, often rotating, that surrounds the participant, and along which they travel. This stage is marked by a 'progressive exclusion of information from outside'.[12] There's also often the experience of bright lights or 'white-darkness'. By this stage, people experience other worlds. These may be realms of the dead, or populated by various types of fantastic beings. They might also feel that they are being transformed, often into animals. Paul Marshall, an expert in mystical experiences, calls stage 3 the 'imaginal' stage.[13]

Trance seems to be a basic, very ancient human capacity. There's plenty of evidence to suggest that prehistoric people experienced trance states. Lewis-Williams suggested that trance might even account for various odd features of prehistoric rock-art. Cave-painting and other rock art by prehistoric people emerged possibly as long as 73,000 years before the present.[14] It often depicts animals like horses, bison, or deer. There are, however, stranger features. Abstract shapes like dots, zigzags, spirals, and grids are common. Some show fantastic quasi-human beings. A number of pieces of rock art also seem to show half-animal, half-human figures. Lewis-Williams suggested that this was because rock art actually depicted trance visions.

This suggestion sparked a ferocious controversy in the rock-art world, which I won't detail here.[15] One problem is that the information that comes down to us from prehistory is always going to be fragmentary. That means that researchers often have to make informed guesses from fragmentary data. And very often prehistorians will differ, sometimes acrimoniously, over very different interpretations of the data. However, Lewis-Williams's main point, that human beings have a basic capacity for extraordinary experiences that seem linked to altered states, seems sound to me. As does the suggestion that this capacity is a very ancient part of human experience. Lewis-Williams suggested that these capacities might have helped spark off both art and religion. And despite the controversy, I find this suggestion broadly plausible.

There's an important limitation to Lewis-Williams model. It is based on the idea of the brain-as-computer and is explicitly called 'the neurophysiological model'. In this model, these experiences can be entirely accounted for in terms of the inner workings of the brain, or as Graham Hancock put it, 'the fevered illusions of disturbed brain chemistry'.[16]

Lewis-Williams does not hesitate to categorise the experiences of trance as hallucinations. This means that they are merely subjective, brain-generated events. Any further significance is imagined, or perhaps imposed by the culture of the person who experiences them. This follows from adopting the perspective of scientific materialism, where altered states are 'nothing but' the brain.[17]

Beyond Neurophysiology?

This is another example of 'the enchanted boundary', where a researcher is able to accept those phenomena which can be accommodated by scientific materialism, and no more. This puts limits on the model. For a start, Lewis-Williams's model does not seem to be able to accommodate *Samadhi* states. This limitation has been addressed by the mystical experience researcher Paul Marshall, who has suggested additional stages for the 'intensified pathway'.[18] Beyond stage 3, Marshall includes substages that he labels 'luminous consciousness' and then 'supreme light'. This is justified by descriptions coming from a vast, global, historical library of mystical literature.

Even the extended 'map' of trance and mystical states isn't perfect.[19] An experience like Ed Mitchell's doesn't seem to fit.

Mitchell didn't pass through anything like hypnogogia or the 'imaginal' state to arrive at what he claimed was *Samadhi*. This is because the map implies that mystical states are generally internal, or more technically *introvertive*. In introvertive mystical experiences, the focus is on an inner state of experience. For example, Christian mystics might describe a union with the mind of God, or Buddhists with ground consciousness. Mitchell's experience was by contrast *extrovertive*. This means that it was focused on the external; in his case, the Earth, the sun, and the stars. Extrovertive mystical experiences can also concern ordinary things. The writer Aldous Huxley described the transformation of a vase of flowers in his office when trying mescaline, another type of psychedelic.[20]

There are other limitations to Lewis-Williams's model. In 2022, David Luke reviewed ten types of transpersonal or parapsychological experiences that have been commonly reported by psychedelics users. These include the perception of additional dimensions, out-of-body and near-death experiences, entity encounters, and different kinds of psi experiences.[21] Some of these experiences might have fully or partially brain-based explanations, like out-of-body experiences, sleep paralysis, or synaethesia — which is where senses seem to mix with one another, so you 'see' sounds as colours, or taste colours. However, even these experiences, which could potentially have brain-based explanations, do seem to have additional dimensions that raise challenges for materialists.

For example, psychedelics users have often reported apparent telepathy, clairvoyance, or even psychokinesis during their 'trips'. According to David Luke, 'Experiences of telepathy were most commonly associated with cannabis, and, to a lesser extent, with LSD, and particularly with MDMA ["ecstasy"], a drug which is characterized by its capacity to induce empathic experiences.'[22] Unfortunately, at this point there have not been too many experimental tests of psi and psychedelics, only 17 published experiments as of 2022. However, Luke did suggest that the results so far represented ‘a promising line of enquiry’.[23]

‘Entity encounters’ are far weirder. Experiencers of trance and psychedelic states have long reported encounters with a dizzying ‘zoo’ of otherworldly beings. The researcher Benny Shanon has even drawn up a fairly lengthy list. Experiencers might encounter mythological beings like gnomes, elves, fairies, or monsters. They might bump into apparently alien beings like the infamous ‘Greys’. Alternatively, they might encounter angelic beings or else semi-divine beings like Jesus or Buddha. If they're unlucky, they might bump into demons, terrifying monsters, or even the angel of death.[24]

Lewis-Williams would no doubt classify every one of these encounters as hallucinations, or totally subjective. And the ‘hallucination’ theory has a lot going for it. I've already mentioned Anil Seth's theory that conscious experience is basically hallucinatory.[25] Although I think that this is very inadequate as a comprehensive theory of consciousness, there's evidence that a good chunk of our experiences are somehow conjured by our brains. What happens normally is that these brain-conjurings are constrained by information coming in through our senses.

Furthermore, in certain circumstances, purely hallucinatory experiences can seem realer-than-real. The psychiatrist Oliver Sacks recounted how he once took a psychedelic and had an entirely hallucinatory visit from some friends which at the time seemed absolutely real.[26] Psychedelics drugs like DMT have also been found in brain-imaging studies to induce a sort of ‘waking-dream’ state, which supports Lewis-Williams's theory of the ‘intensified pathway’ of trance.[27]

At the same time, writing these accounts off as entirely hallucinatory might not be completely satisfactory. Luke himself is ambiguous about their ultimate, possible nature. In his book, he fields several theories about their possible origin. They might indeed be hallucinations of a kind. They might represent otherwise hidden aspects of the user's mind or personality. Or, provocatively, trance and psychedelic states might 'provide access to a true alternate dimension inhabited by independently existing intelligent entities'.[28]

For those willing to dismiss this last possibility, I'd point out that it is the interpretation favoured by many trance-using traditional cultures. If we reject such suggestions without thinking, then we're exhibiting a form of cultural supremacy. This is not only a colonial mentality; it potentially shuts off any deeper understanding. And our culture, with all its serious flaws, seems unlikely to me to hold all the keys to reality.

Still, with Luke's last suggestion, we leave mainstream, materialist science far behind. The possibility of other, alternative dimensions of existence is likely even further beyond the 'enchanted boundary' than psi itself. As such, it's likely to be dismissed out of hand. And yet many users have been driven to consider such conclusions by the sheer intensity and hyper-reality of trance experiences. As far as science goes, recent developments have meant that the experiences themselves have come under significant, new scrutiny.

Psychedelic Revival

People have been using consciousness-altering drugs for many thousands of years. These drugs are known generically as 'hallucinogens', 'entheogens', or 'psychedelics'. There are many different kinds. The Wikipedia page on psychedelics lists hundreds.[29] Many are derived from plants, fungi, or occasionally animals. They include psilocybin from mushrooms, opium from poppies, mescaline from cacti, and THC chemicals in cannabis. Others like DMT, LSD ('acid'), and MDMA ('ecstasy' or 'molly') are synthesised in the laboratory.

Psychedelics have the capacity to alter the consciousness of the user, sometimes drastically. They seem to open a pathway into the 'intensified' trance state that Lewis-Williams described, and can induce powerful mystical states. And they're old. Recently, direct evidence of the psychoactive plant alkaloids ephedrine, atropine, and scopolamine were discovered in a 3000-year-old human hair from a funerary cave in the Western Mediterranean.[30] These chemicals are the signatures of plants like Solanaceae and *Ephedra*. The archaeologists reporting the discovery suggested they were used by shamans 'who were capable of controlling the side-effects of the plant drugs through an ecstasy that made diagnosis or divination possible'.[31]

As for modern times:

In the mid-twentieth century, the lab synthesis of LSD (lysergic acid diethylamide) led to a psychedelic craze that swept America and Europe. This drug was synthesised by Albert Hofmann in the 1940s, who early on managed to give himself a 'bad trip'. After he had dosed himself with the drug, he lay on the couch and found that his

> surroundings had now transformed themselves in more terrifying ways. Everything in the room spun around, and the familiar objects and pieces of furniture assumed grotesque, threatening forms. They were in continuous motion, animated, as if driven by an inner restlessness.[32]

A neighbour who brought him milk was also transformed into a 'malevolent, insidious witch with a coloured mask'.[33]

Despite the potential for bad trips, LSD became popular therapeutically in the 1950s, and by the 1960s was also used recreationally. This psychedelic wave inspired quite a few writers and artists. One of the most influential was Aldous Huxley, best known as the author of the dystopian novel *Brave New World*. Huxley had been deeply interested in mystical experiences since the 1930s, and he ended up taking mescaline, a psychedelic derived from cacti. He described this experience in his book *The Doors of Perception*.[34] In the book, he proposed a theory that the brain acts as a kind of 'reducing valve' or filter, constraining conscious awareness. Psychedelics released this 'reducing valve', allowing access to what he called the mind-at-large. This mind-at-large was a broader consciousness that transcended individual minds and pervaded the universe.

LSD and other psychedelics were embraced by the hippy movement and the counterculture, perhaps in part because of the influence of Huxley's books. The Harvard Professor Tim Leary proselytised the use of the drug, advising young people to 'Turn on, tune in, drop out.'[35] Such sentiments alarmed the authorities in the US, triggering a moral panic. LSD was outlawed in 1967, followed in 1970 by the Federal Controlled Substances Act which outlawed the rest. The authors of the Act claimed that psychedelics were without medical merit, which more or less killed scientific research for decades.[36]

The psychedelics prohibition began to thaw in the 1990s, and by the 2000s several academic and medical institutions had resumed research. This meant that the mystical states of consciousness induced by psychedelic use became once more of mainstream concern. In 2006, a team led by Roland Griffiths at Johns Hopkins Medicine published a study of the mushroom-derived psychedelic, psilocybin. In this study, thirty volunteers were administered psilocybin. The team found the drug was reliably able to trigger 'experiences similar to spontaneously occurring mystical experiences'. They suggested that this would 'allow rigorous scientific investigations of their causes and consequences'.[37] And as commercial interests geared up to make substantial profits off psychedelics, this promise has been vigorously pursued.

The Debate over Cause

Over the last decade, two very different theories have emerged that offer potential explanations for psychedelic effects. These theories are in part responses to a slew of new brain-imaging studies. They also reflect deeper, metaphysical conflicts over the nature of psychedelic experiences.

The first theory is some version of Aldous Huxley's 'reducing valve'. The idea is that the brain somehow filters or transmits, but does not generate, consciousness. Huxley himself was inspired by older writers like the philosopher Henri Bergson, who wrote in the early twentieth century.[38] Similar theories had also been proposed by William James and Frederic Myers, and the basic notion is likely thousands of years old.[39] 'Reducing valve' theory suggests that, under ordinary conditions, the brain acts to block a wider perception of reality. Psychedelics open the 'valve', allowing the perception of a wider reality, beyond the individual person.

According to this theory, the best way to open the 'valve' would be to reduce ordinary brain function. And in 2012, a brain-imaging study was published that seemed to confirm Huxley's theory. The study was conducted at Imperial College, London, led by Robin Carhart-Harris. The team discovered psilocybin-induced decreases in blood flow and connectivity between some key brain regions, instead of the increase the researchers had expected.[40] Specifically, activity was reduced in the default mode network of the brain, which has been speculated to have something to do with the 'ego' or sense of self. At the same time, the participants who took the drug reported profound, meaningful mystical-type experiences. The study authors concluded that the results were compatible 'with Aldous Huxley's "reducing valve" metaphor'.[41]

However, as more studies were done, Robin Carhart-Harris and others began to develop an alternative theory of psychedelics. This was significantly different from the 'reducing valve' theory, and was suggested because the brain-imaging studies had detected an increase in entropy in the brain when people used these drugs. 'Entropy' is a term from physics that refers to disorder and randomness.[42] Carhart-Harris began to formulate a theory that an increase in brain disorder and random functioning was responsible for the 'richness' in conscious experiences that users report.[43] He called this the 'entropic brain theory'. The recent results of a DMT study, and others, seem consistent with the entropic brain theory.[44]

For Carhart-Harris 'entropic brain theory' might also explain the therapeutic effects of psychedelics. These drugs seem to offer potent treatments for things like depression, alcoholism, and drug addiction. One recent Netflix series, *How to Change Your Mind*, presents a number of case studies showing apparently miraculous 'cures' of entrenched psychiatric illness. Carhart-Harris suggested that conditions like depression result in the brain getting stuck in rigid 'grooves'.[45] Psychedelics, by increasing 'brain-noise', somehow kick the brain out of these unhealthy 'grooves'.

So, from a mainstream perspective, there's no need to invoke 'reducing valves' or higher realities to account for the effects of psychedelics. It can all be put down to unusual brain functioning. Michael Pollan, author of a history of the psychedelics revival, puts it in the following way. The highly unusual experiences people report, Pollan claims, get labelled 'spiritual' or 'supernatural' because the 'predictive brain is getting so many error signals that it is forced to develop extravagant new interpretations of an experience that transcends its capacity for understanding'.[46] Pollan is here trying to account for a diverse range of psychedelic experiences in terms of a fairly simple, possibly simplistic, theory of brain function.

For some, 'brain-only' explanations remain unsatisfactory. In a 2018 *Scientific American* blog, Bernardo Kastrup and Ed Kelly pointed to materialistic biases in the worldviews of mainstream psychedelics researchers. They suggested that researchers had become disproportionately fixated on increasing brain activity in the brain scans, and were ignoring significant decreases.

They concluded that 'a formidable chasm still yawns between the extraordinary richness of psychedelic experiences and the modest alterations in brain activity patterns so far observed'.[47] Kastrup and Kelly favoured instead the 'reducing valve' theory.

Proponents of more mainstream approaches quickly kicked back. A writing team including Anil Seth, Robin Carhart-Harris, and skeptic Michael Shermer claimed that Kelly and Kastrup had 'misconstrued' and 'oversimplified' the research. They claimed that Kelly and Kastrup's 'anti-materialistic view of consciousness' had 'nothing to do with the details of the experimental studies'.[48] They suggested that the studies had shown that

> higher signal diversity [i.e., "brain-noise"] indicates a larger repertoire of physical brain states that very plausibly underpin specific aspects of psychedelic experience, such as a blending of the senses, dissolution of the "ego," and hyper-animated imagination. As standard functional networks dissolve and reorganize, so too might our perceptual structuring of the world and self.[49]

They also dismissed the 'metaphysical' dimension of Kelly and Kastrup's critique, claiming that their studies were 'entirely irrelevant to this metaphysical question'.

Bernardo Kastrup replied to this critique on his website.[50] He claimed that the dismissal of 'metaphysics' was itself misleading, because several statements in the critic's article had demonstrated their commitment to materialism, itself a 'metaphysical' position.[51] Kastrup also disputed their interpretation of the psychedelic brain-function studies. He suggested that an increase in 'randomness' in the brain could not account for the highly structured nature and meaning of the experiences. He also highlighted several places where the published data might offer support for the 'reducing valve' theory.

So who's right? It's hard to say. The evidence is changing as new studies appear. However, at the very least, Kastrup and Kelly have shown that alternative interpretations of the data are possible. This is because the brain scan data is quite ambiguous. It's clear that there's a lot going on, brain-wise, in psychedelic states of consciousness. In addition, psychedelic experiences are themselves very complex. Maybe one theory cannot account for all the data.

Perhaps both theories are partly right, or partly wrong. Perhaps, for example, the entropic brain theory better accounts for some of the odd visual effects people experience on LSD, like seeing 'liquid' or 'undulating' surfaces. However, 'psi' or mystical type experiences might still be better accounted for in terms of some kind of 'reducing valve' theory.

Perhaps, too, both theories could be brought together in a broader understanding. Dana Sawyer has suggested that both Carhart-Harris and Kelly's models might indeed be understood in the frame of Aldous Huxley's 'reducing valve' model.[52] For example, increased entropy in the brain might be necessary for the 'dissolution of the "ego"' which in turn could open up the 'reducing valve', allowing intense mystical experiences. Dissolution of the ego is said in quite a few mystical systems, including Patanjali's, to be a prerequisite for perceiving the 'ground' of reality.

Insights into Reality?

The brain scan studies are technically very impressive, offering plenty of new insights into the 'brain on psychedelics'. However, they have a significant weakness. They assume that uncovering the 'underpinning' brain activity 'explains' an experience. This follows from the assumption that the brain scan data is 'objective', but that reports of experiences are 'subjective'. To put it another way: blood flow in the brain is seen as more 'real', or primary, than the person's experience, which is seen as secondary to the physical processes. This makes it very easy to sideline or marginalise the user's lived experience.

This could be a serious shortcoming. Over 100 years ago, William James suggested that it was a mistake to judge experiences by their 'organic origins', or 'nothing but' physiological processes. James rejected this claim of the 'medical materialists', as he called them, for the following reason. If every human experience is mediated by the brain, then

> Scientific theories are organically conditioned just as much as religious emotions are; and if we only knew the facts intimately enough, we should doubtless see "the liver" determining the dicta of the sturdy atheist as decisively as it does those of the Methodist under conviction anxious about his soul [...]. To plead the organic causation of a religious state of mind, then, in refutation of its claim to possess superior spiritual value, is quite illogical and arbitrary [...]. Otherwise none of our thoughts and feelings, not even our scientific doctrines, not even our dis-beliefs, could retain any value as revelations of the truth, for every one of them without exception flows from the state of their possessor's body at the time.[53]

James thought that we should judge states of consciousness by their practical value, and not by their 'organic origins'. So it makes sense to pay a great deal of attention to the practical value of altered states in people's lives. For example, the environmentalist Rachel Petersen found that taking mushrooms for depression 'cured' her of her atheism. This was because at the peak of her experience her 'sense of self dissolved' and she became 'unified with an abiding force that permeated all existence [...] something that felt conscious, vast, benevolent, eternal, peaceful, and furiously important'. As she observes, 'exploding one's worldview is the whole point of these treatments'.[54]

This shift in worldview presents a significant challenge to the 'medical materialists' of our own day. But the voices of experiencers like Rachel are mostly absent from learned papers discussing the brain on psychedelics. Sawyer highlights this omission with a series of questions:

> Why, if the cause of the therapeutic outcomes is simply a reconfigured brain state, do the substances routinely trigger experiences of "God," "the sacred," "ultimate reality," "the fundamental unity of all things," "the transcendent," "divine love," "infinite soul," and other postulates of a transcendent or spiritual nature? Why the consistency? Furthermore, why must these testimonies, so much alike those of the traditional mystics, be disregarded?[55]

The short response to this is because, as previously stated, insights from psychedelics tend to be dismissed as 'subjective', unlike the outputs of brain scanners and AI programs. However, I'm not sure that those gunning for a purely materialistic explanation have the right to summarily dismiss the experiences of people like Petersen. This does not mean that we have to take her conclusions at face value, but it does mean that such insights deserve serious consideration. There exists the possibility that mystical states might allow insights into reality that are simply not available in everyday states of consciousness. Mystical experience researcher Paul Marshall recently addressed this question:

> Does mystical experience give access to reality? At the very least, [Sarah Ritchie] is surely justified to say [...] that "it is at least plausible to suggest that mystical psychedelic experiences are able to yield insights that *may* be rooted in fundamental reality, even Ultimate Reality".[56]

The idea that mystical states can somehow allow access to 'Ultimate Reality' is deeply provocative. It could imply that consciousness is somehow fundamental to reality itself. But is this sort of claim even remotely plausible? Might conventional explanations, after all, be enough? And if not, how does the evidence, or alleged evidence, for psi fit in? And how about the wider range of exceptional human experiences? Also, how might conventional understandings of brain function, biology, and evolution be reconciled with what currently seems to lie beyond? These tough issues are the subject of the next chapter.

Chapter 8

In Theory

In April 2000, I made a clandestine visit to the Perrott-Warwick conference at Trinity College, Cambridge. The Perrott-Warwick fund was a bequest to the College, made in 1937, specifically for the investigation of 'mental or physical phenomena' that seemed to 'suggest the existence of supernormal powers of cognition' or else 'the persistence of the human mind after bodily death'.[1] The fund had been providing support for parapsychology ever since. The 2000 meeting consisted of about 50 participants including a number of prominent figures in the field of parapsychology as well as some leading skeptics and other notable scientists, some of whom were world leaders in their field.

My visit was clandestine because at the time I was completing a conventional doctoral programme at the Centre for Computational Neuroscience and Robotics in Sussex. The ethos of the Centre was very mainstream cognitive science. Several of the members of the Centre were aware of my interest in parapsychology, which had raised some eyebrows. But a friend of mine had offered to get me an invite, and curiosity had overridden the faint disapproval of my peers.

Trinity College, Cambridge, is about as prestigious as they come. Founded by Henry VIII in 1546, it has extensive gardens and is based around the Great Court, one of the largest enclosed courtyards in Europe. It was attended by the current King, Charles III, and by numerous Nobel Prize winners. The theme of the conference was 'Rational Approaches to the Paranormal', although I was to witness some fairly emotional, perhaps irrational, outbursts while attending.[2]

Over 20 years later, I have a number of fond memories of that conference. It was attended by several people who've remained good friends. It also featured a number of well-known figures in parapsychology who are no longer with us. This included Ian Stevenson, an American psychiatrist and researcher who researched children's apparent past-life memories. Also present was Professor Robert 'Bob' Morris, who at the time sat in the Koestler chair of parapsychology at Edinburgh University. Morris was in part responsible for establishing a network of parapsychology scholars and students not just in Edinburgh but throughout the UK. He was also keen to establish bridges with mainstream psychology and other branches of science.

It was also interesting to meet and converse with a number of leading UK skeptics. I had some fairly extensive conversations with Susan Blackmore and Richard Wiseman, who were both prominent in the media at the time. These conversations were friendly, although this was not true of all the skeptical talks. One psychologist who was also a vigorous skeptic began actually shouting during his talk. Another skeptical psychologist began to shout after a parapsychology-friendly talk on consciousness by a physicist. This is something that I've encountered occasionally since: a sense of sheer outrage at the possibility of something like psychic phenomena being true.

This sense of outrage was also present, but in a milder form, at a lecture given by Basil Hiley, a leading physicist. Hiley seemed happy to talk about the most esoteric dimensions of physics and even consciousness. However, the topic of psi triggered an uneasy frown. I don't think that Hiley was rejecting the idea of psi, per se. It was more that he didn't quite see how to fit it within his view of the world. He was not the only one to express this sort of sentiment. It was as if the scientists had been completing a mostly blue and green world map jigsaw puzzle and had discovered a large, red and misshapen piece in the bottom of the box.

This chapter explores how the 'missing piece' of psi might fit within a perhaps expanded scientific theory of the world. I'll cover three possibilities:

1. ***Psi does not exist.*** 'Anomalous' and 'exceptional' human experiences can be explained entirely in terms of conventional psychology, neuroscience, and culture.
2. ***Psi does exist,*** but 'anomalous' and 'exceptional' human experiences can be explained mostly in terms of conventional psychology, neuroscience, and culture.
3. ***Psi does exist,*** and exceptional human experiences starting with consciousness need a radical new world-picture and a radical new theory to be understood.

Let's begin with the first possibility.

Psi Does Not Exist. 'Anomalous' and 'Exceptional' Human Experiences Can Be Explained Entirely in Terms of Conventional Psychology, Neuroscience, and Culture.

This has also been called the 'smoke without fire' theory.[3] From this perspective, psi is essentially illusory. The same goes for out-of-body experiences, near-death experiences, apparition encounters, and the rest. This is the position advocated by the skeptics. It's powerfully expressed by Susan Blackmore and Emily Troscianko in their textbook on consciousness. There is even a practice exercise titled 'Living Without Psi'. They suggest that 'The possibility of ESP is comforting', but that for the exercise people should try living 'without such comfort'.[4] The idea is to set aside any pre-existing belief, or disbelief, in paranormal phenomena and see what life is like. The key question is: 'Do we live better for a belief in the supernatural?'

This is a good question, and I suggest that the reader tries the exercise for themselves.[5]

So why remain skeptical about psi? In 2003, the skeptical psychologist James Alcock laid out a number of 'Reasons to Remain Doubtful'.[6] Alcock felt that paranormal experiences could be accounted for as the 'product of normal but misunderstood brain function'.[7] This response is pretty much identical to that of Brian Cox, who'd stated that psychic experiences were in fact people 'processing stuff' in their 'remarkable brains'. Alcock acknowledged that 'these experiences appear to be relatively common and are often very striking' and that they merited 'study in their own right'.[8] But he rejected the idea that any of those 'striking' experiences might reflect unknown processes like telepathy or genuine precognition.

But what sorts of psychology and neuroscience explanations might help account for paranormal experiences? 'Cognitive biases' are one plausible source. Many experiences of apparent telepathy and precognition, after all, rely upon apparently extraordinary coincidence. You might be thinking of a friend when they ring you on your phone. Or you might dream of a plane crash, turn on the news in the morning and be confronted by the news of a real crash. These might be explained in psychological terms by what is called 'magical ideation', or the tendency to link events that are not in fact associated.[9] People who report psi experiences, the argument goes, have a tendency to seek patterns in random events. Random coincidences merely trigger this tendency. So skeptics like Chris French insist that a number of ordinary 'cognitive factors' underlie many apparently paranormal accounts.[10]

What about ghosts or other apparitional experiences? These can be explained entirely as subjective hallucinations. The psychologist Richard Bentall notes that hallucinations, or the perception of things that aren't there, are actually very commonly experienced by psychologically normal people.[11] Blackmore and Troscianko lean heavily on the idea that hallucinations and 'imagination' can explain a wide range of apparently paranormal experiences.[12] For example, Blackmore has presented quite compelling evidence that out-of-body experiences are related to a disruption in body image and tend to be experienced by people with a strong visual imagination.[13]

But what about events that seem to defy even unusual psychological explanations? One example, in the Enfield Poltergeist case in 1977, a police officer and seven other people saw a chair begin to wobble from side to side then slide 3 or 4 ft across the room. The police officer reported that she 'checked the chair but could find nothing to explain how it had moved'.[14] Later on in the case, an investigator, alone in the kitchen, saw a teapot 'rock back and forth for about seven seconds'.[15] There are actually quite a large number of cases like this in the archives of parapsychology. Cases where eyewitnesses report things that seem to defy common-sense explanations.

No problem, the skeptics say. The Enfield case specifically did not impress everyone. Quite a few experts who attended saw nothing paranormal. A leading critic of the case, Anita Gregory, claimed that the leading investigators, Maurice Grosse and Guy Lyon Playfair, had a tendency to interpret mundane events as paranormal. She even claimed that the police officer had been 'coached' towards a paranormal interpretation of the chair incident by the investigators.[16] These sharply contrasting perspectives are typical of the controversies surrounding paranormal phenomena.

More generally, there are quite a few studies that show that eyewitness testimony is poor. The classic experiments on this were done by a psychologist named Elizabeth Loftus.[17] She found that people made multiple errors in recalling staged events. People also have a tendency to confabulate, and their brains can even manufacture totally false memories. For example, David Hall, Susan McFeaters, and Loftus discovered a number of alterations in eyewitness accounts of various 'unusual and unexpected events'.[18] The British skeptic Chris French also often points to the fallibility of memory in order to cast doubt on paranormal reports.[19] The implication is that eyewitness testimony and human memory are so unreliable that they are worthless as evidence for truly paranormal events.

Although it's important to acknowledge potential weaknesses in eyewitness testimony, for me, the skeptical arguments are overstated. Eyewitness testimony is not always as fallible as critics imply. A recent experiment testing adults and children's tendency to produce 'false memories' found that children were less prone to confabulate than previously believed. In fact, in some conditions, 'Children can provide not only highly accurate accounts, and are ofttimes less vulnerable to the formation of spontaneous memory errors, but also false memories, as the result of external suggestion.'[20] What the team found was that context mattered significantly in terms of whether children and adults recalled something accurately or not.

Other skeptical arguments about spontaneous paranormal phenomena can also be at least partly addressed. In 2010, in a book titled *Randi's Prize*, the journalist Robert McLuhan looked at skeptical arguments in detail. He examined several areas of controversy within parapsychology, including arguments over poltergeist cases, mediumship, and experimental work. In each case he found significant weaknesses in the claims of the skeptics. Often psychological analyses 'didn't connect in any obvious way with the experiences that interest psychical researchers'. Instead, in the case of spontaneous cases, 'what one gets is quite a lot of insightful psychological material about faulty reasoning, with episodes from debunking literature tacked on afterwards, but no indication of what the two have to do with each other'.[21]

We've already looked at the strengths and weaknesses of skeptical arguments for experimental evidence. In his 2003 article, James Alcock emphasised the failure to achieve replication and pushed a number of generic arguments (lack of progress, flaws in experiments, etc.) that have been repeatedly addressed by parapsychologists. Alcock's more recent refusal to look at the empirical evidence at all, simply because psi 'cannot' be true, seems to highlight the weakness of the skeptical case.[22]

Psi Does Exist, but 'Anomalous' and 'Exceptional' Human Experiences Can Be Explained Mostly in Terms of Conventional Psychology, Neuroscience, and Culture

The 'smoke without fire' theory has strengths but also significant weaknesses. It takes mainstream assumptions about mind, consciousness, and brain for granted. So psi phenomena are held to be not just unlikely but 'impossible'. Sometimes psi is held to be impossible because it violates the 'laws of physics'. So for example, psi routinely seems to violate the 'common-sense' assumptions of local cause and effect. It should not be possible to have direct knowledge of distant places, of the future, or of the contents of someone else's mind without some meditating chain of cause and effect. (e.g., seeing the events on TV or asking someone for their thoughts). The locality of cause and effect is one of the basic assumptions of classical physics.

However, psi seems to demonstrate nonlocality. This is a term borrowed from quantum physics. In quantum physics, nonlocality refers to an odd effect of entangled particles. It originated with what is known as the 'Einstein Podolsky Rosen' (EPR) paradox.[23] The EPR paradox seemed to imply that the spin of two entangled particles would be equal and opposite to one another, no matter how far apart they were in the universe. The physicists Einstein, Boris Podolsky, and Nathan Rosen introduced this paradox in order to demonstrate the absurdity of quantum theory, but nonlocality has by now been demonstrated in many laboratories around the world. The consensus is now that the world as described by quantum theory is nonlocal. So in one quite narrow technical sense, quantum theory also violates the common-sense assumption of local cause and effect.

It seems to me that the assumption of locality is a major skeptical stumbling block for the acceptance of psi. But if you allow some form of nonlocal, time-and-space-defying connection between minds and the universe, then psi not only becomes possible, it seems debatably likely. And it's tempting to begin to relate nonlocality in physics with the apparent nonlocality exhibited in psi experiences. This has problems, of course. One is that the kind of nonlocality described by the EPR paradox forbids the transfer of information.[24] Another is that skeptics have often rejected claims that psi might be relatable to quantum theory.[25] However, Dean Radin has offered the counter-suggestion that 'What quantum mechanics does provide, whereas classical mechanics does not, is evidence that the physical world is compatible with psychic phenomena.'[26]

Still, the assumption of locality represents a line in the sand beyond which skeptics refuse to cross. It's a line that prevents not just skeptics but also psychologists, neuroscientists, and philosophers from taking psi seriously. I do wonder whether a significant degree of the resistance for psi would collapse if nonlocality became part of the basic background assumptions in biology and neuroscience. If there was such an assumption shift, then perhaps psi information transfer would be accepted as a normal, biological, consciousness-related process. And this suggestion may not be as far-fetched as it might at first appear.

Nonlocality aside, one of the main things driving skepticism is the assumption that parapsychology is by its nature totally incompatible with mainstream neuroscience. In a 2023 invited commentary in the *Journal of Scientific Exploration*, the cognitive neuroscientist David Acunzo suggested that the idea of psi 'goes against the intuitive worldview of (most) cognitive neuroscientists'.[27] This comes back to a view of the brain as fundamentally machine-like or computational. So, for many neuroscientists, 'there is naturally little room for this phenomenon' because psi 'implies the existence of unknown fields, forces, and interactions, an unknown way to reach and receive that information across space — and time if one wants to account for precognition'.[28] Despite this, Acunzo suggested that neuroimaging and other techniques from the mainstream might be very useful in parapsychology research.

One school of thought in parapsychology by and large adopts the approach suggested by Acunzo. This school accepts most of the findings of neuroscience. They're happy with the efforts to reduce human experience to brain function. They accept that many, probably most, allegedly paranormal phenomena are the products of faulty observation, deception, and self-deception. They reject any talk of spirituality or the soul. Some even embrace atheism. They also think that at least some psi phenomena are real.

The veteran parapsychologist Richard Broughton more or less takes this position. Amongst other things, he was director of the Institute for Parapsychology in Durham, North Carolina, from 1981 to 2000. He is interested in how psi night work in the brain, and also in the context of evolution.[29] His approach is centred on neuroscience, and he rejects the idea that mind and brain might be totally different things:

> For me, the dualist view—that consciousness exists separately from the brain—was simply one rather general hypothesis that could be entertained to explain the data, but not one that leant itself to my goal of understanding how it all worked [...]. The developments in neuroscience that have [recently] unfolded [...] have been profound and exciting, and, in my opinion, increasingly render the dualist hypothesis at least unnecessary, if not untenable.[30]

Essentially, Broughton thinks that psi should be understood in the context of very mainstream neuroscience and evolutionary theory. He thinks that this shift is essential if the findings of parapsychology are ever to be accepted within the broader body of science. However, he remains willing to abandon that stance if there is evidence that points strongly enough in a different direction. He remains open-minded about such things as poltergeists, large-scale psychokinesis, and even potential reincarnation evidence.

Researchers behind the US intelligence programme Stargate have adopted a similar stance to Broughton. Dr Sonali Marwaha stated that, from the first, those working in Project Stargate and following programmes took a physicalist (or materialist) position on remote viewing. The approach was 'primarily a physics, engineering, and cognitive sciences approach'. The team made 'absolutely *no* mention of terms such as consciousness [...] non-local consciousness, spirituality, dualism, or religion' in their reports.[31] They framed their work in terms that were as close to mainstream work as possible.

The team concluded that informational psi probably existed, but that this was probably limited to a 'trans-temporal' acquisition of information.[32] In other words, they thought that the remote viewers were sensing the future only.[33] They also thought that the phenomenon 'resisted training' and also acknowledged problems with replicating results.[34] But the bottom line was that they thought that the only real psi effect was precognition, or information transfer from the future.

However, there are reasons for doubting that this kind of restrictive view can account for all of the data. Parapsychologist David Vernon has suggested that the evidence that precognition underlies other psi processes is currently 'scant'.[35] He also highlights evidence suggesting that the effect sizes for telepathy seem higher than for precognition, perhaps suggesting different processes. In addition, I'd question the team's blanket rejection of nonlocality as a concept. And if a kind of anomalous information transfer can occur across time, I do not see why it cannot occur across space.

Jim Carpenter's 'first sight' model represents another attempt to show that psi might not be exceptional or 'miraculous'.[36] First sight builds upon the insight that many things have to happen in the body and brain before we experience any emotion or thought. This activity is known as preconscious. There also exists a form of perception called *subliminal perception*. Subliminal perception occurs when we experience something like an image or sound too fleetingly to register consciously. Studies have shown that such fleeting perceptions can still have an influence on our emotional responses and on our decisions. Subliminal perception has been studied since the 1950s, and with the publication in 1957 of Vance Packard's *The Hidden Persuaders*. It has often been used in advertising.[37]

Carpenter suggests that psi works like subliminal perception. In other words, it's not available to conscious experience but works as a subliminal 'prime' that influences thoughts, feelings, and emotions. He suggests that informational psi, like telepathy or clairvoyance, operates like an 'input' in this way. Psychokinesis operates in the opposite way, reaching outwards as 'unwitting expressions of unconscious mental processes'.[38] Carpenter proposes that psi operates continuously, all the time. In fact, he proposes that psi forms the first, preconscious stage of all conscious experience. So psi is not rare but a normal, if subtle, part of human functioning.

This model rests on two assumptions. The first is that 'organisms are psychologically unbounded. They transact with reality in an unconscious way beyond their physical boundaries.'[39] This connection is not an impersonal, mechanical one; Carpenter conceives an 'extended universe of meaning of indefinite extent in space and time'.[40] This means that he breaks with two standard assumptions about how the universe works (standard for biology or neuroscience, but not necessarily for physics). These are the assumptions of local action only and the assumption that the basic processes in the universe are impersonal and mechanical. Carpenter suggests that meaning can be viewed as somehow inherent and something with which humans transact on a continual basis.

The second assumption is that all experiences are 'constituted of unconscious psychological processes' that are 'carried out purposely on multiple sources of information'. They are impersonal in the sense that they are unconscious, but 'they are not impersonal or mechanical'.[41] This information comes from many different places: 'including nonlocal information [psi], sensation, memory, imagination, goals and values'.[42] Carpenter thinks that things like ordinary sensory information will be used preferentially over psi information, which is subtle and tends to hide in the background of our choices and experiences. The unconscious nature of psi is why it's rarely noticed in the everyday and only becomes apparent in exceptional circumstances; Carpenter compares these moments with 'lightning bolts'.

In the book, Carpenter also works out a number of implications for his theory in terms of psychology. He thinks that psi operates in a comparable way to memory. He also thinks that the somewhat spotty experimental results can be explained in terms of ordinary psychological dynamics. He compares psi with other preconscious psychological processes, including ambiguous stimuli, compares psi with memory, and compares it with creativity. Carpenter claims that his model can explain the interactions between psi performance and personality types (e.g., extroverts and introverts). He also thinks that it might explain so-called 'experimenter' and 'decline effects'. In other words, he treats psi like a jigsaw piece that actually does 'fit' within the larger puzzle of human personality and consciousness.

In my view, first sight partly answers the skeptical objection that parapsychology lacks a theory. First sight seems to me a comprehensive and reasonably plausible account of how psi might operate, and also how it might relate to 'ordinary' psychology, memory, personality, and consciousness. This is crucial because it implies that psi is not some weird 'superpower' or 'force' that is totally separate from other human faculties. Instead, it's an integral part of the basic constitution of human beings.

The 'mainstream neuroscience plus psi position' has many strengths, but for some it's not radical enough. The reason is that those who advocate this position also seem to accept that neuroscience will eventually understand consciousness solely in terms of brain function. There are those who think that such reconciliation is implausible. They point to the wider spectrum of exceptional human experiences to bolster their arguments. We turn, then, to the radical proposal: that psi is not a 'jigsaw piece' to be fitted into otherwise mainstream knowledge, but the thin end of a wedge that might allow a radical worldview overhaul.

Psi Does Exist and Exceptional Human Experiences Starting with Consciousness Need a Radical New World-Picture and a Radical New Theory to Be Understood

In 1998, Michael Murphy, one of the founders of Esalen, organised the first meetings of what became known as the 'Sursem' group.[43] The Esalen Institute was founded at Big Sur in California in 1962, and advertises itself as 'the epicentre of the Human Potential Movement'. The goal of Esalen was and is to explore 'new ideas around creativity and the brain, body work, spirituality, leadership', and a number of other esoteric topics.[44] The 'Sursem' group was a 'long running intellectual fellowship' convened to study the question of whether any portion of human personality survived bodily death.[45] ('Sursem' is an abbreviation of 'Survival Seminar'.)

The group soon realised that their task needed to be significantly broader and widened the breadth of their research to include the broad landscape of exceptional human experiences. Their aim was firstly to 'assemble in one place many lines of peer-reviewed evidence demonstrating empirically the inadequacy of conventional physicalism'.[46] Their second task was to 'seek some better conceptual framework to take its place'.[47] Their work is ongoing and has resulted in the publication of three large volumes that summarise their findings. The first book, *Irreducible Mind: Toward a Psychology for the 21st Century,* brought together a wide range of evidence that seemed to show that the human mind could not be fully understood in terms of conventional neuroscience and biology.[48] The second and third volumes, *Beyond Physicalism: Toward Reconciliation of Science and Spirituality* and *Consciousness Unbound: Liberating Mind from the Tyranny of Materialism,* presented a range of different possible theories and expanded worldviews that the multiple authors felt might explain the data.[49]

What evidence did the team feel suggest the inadequacy of materialistic or physicalistic approaches? One of the primary members of the Sursem group, Ed Kelly, offered an initial list at the start of *Irreducible Mind* and in the subsequent volume, *Beyond Physicalism*. Psi phenomena are included and also 'extreme psychophysical influences' such as stigmatics' capacity to produce the bleeding wounds of Christ on their bodies. There were also cases of human minds exhibiting what Kelly termed 'informational capacity, precision, and depth', as well as exceptional memory, 'psychological automatisms and secondary centres of consciousness', genius-level creativity, and finally, mystical experience.[50]

Materialism does seem to founder with respect to mystical experience. This seems especially true if you include insights from the experiencers of ecstatic and mystical states. Previously, I mentioned the account of environmentalist Rachel Petersen, who experienced a complete transformation of herself and the world after taking psilocybin for depression. I would repeat her conclusion that 'exploding one's worldview is the whole point'.[51] The noted early twentieth-century scholar of mysticism Evelyn Underhill suggested that mystical experiences were better understood from a worldview of 'vitalism' as opposed to materialism. Materialism states that we live in a mechanical universe, grinding out physical laws. 'Vitalistic' philosophies, by contrast, emphasise 'Not law but aliveness, incalculable and indomitable [...]. Vitalists see the whole Cosmos, the physical and spiritual worlds, as instinct with initiative and spontaneity: as above all things free.'[52] This contrasts notably with a worldview that insists that the universe is purposeless and that we ourselves are bound by natural laws, lacking any kind of free agency.

A defender of mainstream approaches might reply that mystical experiences are subjective and that they cannot be used to build any kind of scientific theory. That the only 'objective' evidence is that gathered by brain scans or recorded in statistical tests in controlled experiments. And there's truth to this objection. William James himself felt that mystical experiences were not coercive for people who had not experienced them.[53] He also said that non-experiencers must accept that they were coercive for the experiencer. This means that simply asserting that a user is deluded or mistaken will not do. So, despite the difficulties, the possibility that mystical experiences reveal a deeper and broader aspect of reality beyond ordinary rational-analytical consciousness remains a live one.

This throws up difficult questions about what kind of evidence is acceptable when you're building scientific theories. Many skeptics seem to take the perspective of the hard sciences. This means that they will only accept hard-science kinds of evidence, primarily evidence gathered in a laboratory under strictly controlled conditions. Kelly dubbed this focus 'methodolatory', which is 'the methodological face of scientism'. He suggested that 'Laboratory experimentation certainly does not exhaust the means of obtaining valid and important information.'[54]

Quite a few things in this world cannot really be studied or 'proved' in hard-science terms. This is especially true when investigating human phenomena. This means that much of the material with which we are dealing must really be examined from the perspective of the social sciences. The best case studies of spontaneous, apparently paranormal phenomena cannot and should not be dismissed as 'mere anecdotes'. They exist as part of a gradation of evidence that ranges from accounts to case studies to controlled experiments.

For example, some of the best evidence suggestive of the survival of human personality beyond death exists as case studies. This includes quite a few cases where very young children seem to remember often accurate memories of past lives. The researcher Ian Stevenson and others collected over 2500 cases of the reincarnation type.[55] This evidence is included by the Sursem group as part of the case against physicalism.

What sort of theory did the team feel might explain the wide range of exceptional human experiences, as well as consciousness itself? In short, some version of the idea that the brain filters, transmits, or permits consciousness as opposed to generating it. *Irreducible Mind* also sought to rehabilitate a particular version of this theory, the subliminal mind, propounded by Frederic Myers. Myers, one of the founders of the Society for Psychical Research in the UK proposed that waking consciousness was a mere fragment of a far vaster consciousness that existed beyond the confines of the body and brain.[56] He also suggested that this consciousness had an 'infrared' and 'ultraviolet' portion. The infrared portion existed 'below' waking consciousness and consisted of instincts, complexes, and neuroses. However, the 'ultraviolet' portion, which Myers saw as emergent, included the capacity for psi, intense mystical experience, and was also responsible for the products of genius.

The totality of this sort of theory takes us far beyond mainstream materialism. The question is whether such a radical move is justified. The answer to this question very much depends upon how you interpret the evidence gathered by the Sursem team as a whole, and whether their overall interpretation of this accumulated evidence is valid. It also depends very heavily on what sorts of things you are willing to accept as evidence in the first place. This is no doubt also debatable, and such debates are beyond the scope of this little book. At the very least the Sursem team have drawn attention to a vast quantity of exceptional human experiences that have often been marginalised, minimised, or ignored in the name of keeping the human sciences 'pure'. This kind of restriction seems increasingly indefensible.

Where Do We Go from Here?

So there do exist potential alternatives to mainstream approaches. And the evidence for psi and possibly more extreme exceptional human experiences seems to compel a reconsideration of those alternatives. This reconsideration is urgent for both theoretical and very practical reasons. We will see next that there's evidence that a stubborn adherence to the 'dogmas' of scientific materialism is now causing more harm than good. This adherence, in turn, is having a deleterious effect upon how mainstream culture approaches the multifarious challenges of the future.

Chapter 9

A Dysfunctional Future?

In 2002, a freshman from the California suburb of Palo Alto stepped in front of a train.[1] This was the start of a chain of teenage suicides which continued through the 2000s and 2010s. Writer and historian Malcolm Harris was at the time attending school in Ohlone Elementary in Palo Alto. The suicides formed a disturbing background to his school life. The official explanation was that the deaths were part of a 'suicide cluster', essentially a form of mass hysteria. This was an explanation that Harris and his contemporaries never found very convincing.

Palo Alto is 'nice', Harris states on the first page of his history. 'The weather is temperate; the people are educated, rich, healthy, innovative.'[2] It's located within Silicon Valley, one epicentre of the techno-utopian revolution now sweeping the planet. It's the residence of tech investors and resident coders, who helped build the information technology and AI revolutions. Palo Alto embodies the myth that, as Harris puts it, 'hard work and talent allowed some people to change the world single-handedly'.[3]

Growing up, Harris found himself preoccupied by what might be called Palo Alto's shadow.[4] The distress of the kids was only the tip of a rather unsettling iceberg. He discovered that Palo Alto was built upon conquest. The indigenous peoples succumbed partly to diseases introduced by European colonists. An eliminationist and displacement programme took care of the rest. As Harris puts it 'the California genocide was a bottom-up, settler-led process'.[5] Palo Alto was the site of the 1849 Gold Rush, boom time for entrepreneurs and free-traders. Aggressive capitalism was enabled by technological support. Harris reports that 'California engineers became the [...] shock troops of global enclosure, drawing lines where others were forced to follow.'[6]

Palo Alto exemplifies a number of broader patterns of world history, starting with the extermination and displacement of indigenous people as part of aggressive colonisation. In Palo Alto, as elsewhere, the alliance of science and technical expertise with extractive capital is part of this broader pattern. Like it or not, science is one local kind of knowing that has been spread by conquest, empire-building, and latterly, capital-driven globalisation. This process has successively displaced other ways of knowing about the world. This displacement has often been justified by the claim that science is somehow 'objective', 'neutral', and 'universal'.[7]

I say this as someone who's very dedicated to the practice of science, and who's endlessly fascinated and awed by the tidal wave of discoveries at the cutting-edge of twenty-first-century science. There's no denying that the sciences have been immensely successful in ascertaining truths about the world in which we live. Secondly, I do think that these discoveries are genuine achievements. Humanity is surely in a better place knowing that the Earth goes round the sun, that water is made of hydrogen and oxygen, that viruses and bacteria cause disease, that evolution occurs, and that the continents move. The technical achievements of applied science also need to be acknowledged: the computer on which I write these words is only possible because of Turing and Church's theoretical work on computation, and because of the quantum theory formulated by a group of geniuses in the early twentieth century.

At the same time, there's always a price to be paid when one way of knowing displaces others; the world in which we live is a complex, mysterious, and maybe even unfathomable place; the theories and models of the world created by any human system, including twenty-first-century science, seem unlikely to me to be able to cover everything; there are always going to be things that are ignored, rationalised, and displaced because they don't 'fit'; and we've seen throughout this book that the 'hidden events' of psychic and mystical experience have been repeatedly displaced and excluded by the practice of technoscience, pushed beyond the 'enchanted boundary', and not really taken seriously.

This sort of exclusion is especially a problem when 'science' degenerates into scientism. Scientism is the belief that science is omnicompetent and that all other forms of knowing are deluded or wrong. The philosopher Mary Midgley described scientism as the 'over praising' of 'one sort of very abstract knowledge', and the 'assumption that this knowledge is a 'victor' that has rendered alternatives irrelevant'.[8] There's a case to be made that we live in a culture with an unofficial belief system of scientism. This is allied to a worship of technology as an end in itself. This belief system aims for full-spectrum dominance, excluding alternatives, resulting in a very particular vision of the future.

Technotopians Rule?

Silicon Valley is perhaps the most visible representative of what have been called 'technotopian futures', a term coined by futurist Jennifer Gidley. According to Gidley, technotopian visions are based upon a 'mechanistic, behaviourist' model of human beings, with 'a thin cybernetic view of intelligence'.[9] Gidley also describes this kind of future vision as 'dehumanising, scientistic, and atomistic'.[10]

According to tech commentator Douglas Rushkoff, this hard-line materialist, scientistic approach is frequently taken as read by the elites of Silicon Valley. In his book *Survival of the Richest*, he reports that the 'new atheists' and materialism advocates Daniel Dennett, Richard Dawkins, and champion of the capital-friendly version of the Enlightenment, Stephen Pinker, have all been feted by this community.[11] This association by itself does not invalidate materialism. But equally, a worldview that reduces the cosmos to purposeless matter in motion can enable some very toxic practices because it can easily be interpreted as amoral.

Rushkoff links Dawkins's hard-line materialist views with what he calls 'the Mindset'. 'The Mindset', which Rushkoff also calls 'Silicon Valley Escapism', is

> based in a staunchly atheistic and materialistic scientism, a faith in technology to solve problems, an adherence to biases of digital code, an understanding of human relationships as market phenomena, a fear of nature and women [...] and an urge to neutralise the unknown by dominating and deanimating it.[12]

'The Mindset' is ruthlessly extractive. It exists for the purpose of accumulating technological power and wealth. It appropriates science, or a veneer of science, for its own ends. Rushkoff: 'forcibly removed from the greater contexts of meaning and morality, science easily falls into the service of domination and control'.[13]

'The Mindset' enables a programme of technological development for profit that relies upon a very impoverished, robotic view of human beings. Rushkoff outlines this in an earlier book, *Team Human*. Our world is seen as 'computational. Everything is data, humans are processors.'[14] Rushkoff dubs this tendency *mechanomorphism*, or the tendency to see living things as machines. The final aim, claims Rushkoff, is to transcend humanity and become a machine. This is the ultimate goal of the scientistic worldview that is now the ideological engine of a good chunk of global capitalism.

Technotopianism often presents itself in visionary or revolutionary terms. One recent example, an article in *Psychology Today* heralding the dawn of the 'cognitive age'. The author assures us that technological revolutions have periodically 'fundamentally altered the trajectory of civilisation'.[15] We're told that the Fifth Industrial Revolution will be powered by AI and will transform the way we think about minds. Out will go Cartesian dualism, in will come a new way of thinking about thinking, inspired by the incorporation of sensory data into 'traditionally cognitive machines'.[16] This shift is depicted as inevitable and unstoppable.

The creed has a number of different forms. Recently, a philosophy called 'longtermism' has been doing the rounds, especially promoted by the philosopher William MacAskill.[17] Longtermism is currently enjoying considerable influence, thanks to promotion and financial support by billionaires like Elon Musk.[18] Longtermism presents a vision of the future as endless growth into the universe. It's techno-capitalism to the max. In this future, humans will no longer be flesh and blood but digital and will exist in their trillions, scattered throughout space. The Earth, meanwhile, will have been totally subjugated to human control and its ecology 'tamed'. Wild animals will have been for the most part wiped out as MacAskill rates their value as far less than that of human beings. This is sold as a desirable, even necessary, vision of the human future.

But there are signs that technotopianism in its various forms has become more of a problem than a solution for our various global, human, and ecological problems. The vision of human beings, nature, and the universe presented in technotopian vision is highly impoverished. An adoption of something like Rushkoff's 'Mindset' seems to cause a kind of tunnel vision which is already impacting in very negative ways upon human life and on the natural world.

How is this relevant to the question of psychic phenomena and mystical experience? Those advocating technotopian futures tend to take it for granted that the universe and human lives are ultimately 'pointless'.[19] This latter dogma clashes with mystical experiences that powerfully suggest to the experiencers a universe pregnant with meaning, purpose, and vital force. From the point of view of a technotopian, these 'insights' can only be brain-generated delusions. It's also taken for granted that psychic phenomena are illusory and so have nothing to offer in terms of potential insights into the way the world and consciousness might work.

Therefore mystical and psychic experience cannot be part of the technotopian vision. This dimension of human experience becomes excluded and obscured. This may be a crucial omission because such experiences suggest that every human has deep, intimate connections with other people, nature, and the universe as a whole. In the absence of this insight of deep connection, we're left with a philosophy and vision of disconnection. And this fundamental sense of disconnection is making the current 'polycrisis' far worse. This can be seen by looking a little closer at two distinct facets of that polycrisis. One facet is humanitarian, the other ecological.

The Human Crisis

Gabor and Daniel Maté's book *The Myth of Normal* provides a comprehensive survey of all the ways that industrial consumer societies traumatise and alienate their own citizens.[20] First, the Matés outline what they see as the basic human needs 'encoded in our biology'. It turns out that we're not selfish, atomised individuals. Instead we have a fundamental need for mutual support, for personal connection and belonging. Our nervous systems also come attuned with certain expectations. Trauma happens when those expectations are not met. We need social conditions that nurture us and allow us to reach our natural potentials. This is especially true of children.

The Matés lay out multiple roads to distress. Dehumanised medicine, common parenting practices, education, and hyper-capitalism act as petri dishes that cultivate widespread trauma. The result is routinely distressed, unhealthy, unhappy individuals. Racial, class, and gender differences allied with significant economic inequality have a high impact on health outcomes. Institutions like politics and big business tend to encourage 'sociopathy' and sociopathic practices.

The Matés point to 'neuromarketing' as an example of this. 'Neuromarketing' is the practice of using neuroscience and psychology to sell products.[21] They detail how it was used to sell addictive junk foods to populations and therefore contribute to widespread poor health. This has harmed whole populations, and they conclude that the use of specialist science to maximise calorie intake in customers surely qualifies as 'sociopathic'.

This crisis extends to politics. One big problem is that our political institutions seem to enable psychologically damaged people. The Matés name several notable examples, and they're far from the only ones to have seen this. The UK psychologist Steve Taylor has written about what he calls 'hyper-disconnected' people, a condition that's endemic in our culture.[22] These are people who are psychologically and emotionally isolated from everyone else, whom they tend to see as manipulable objects. Hyper-disconnected people tend to damage themselves, other people and the world.

The Matés ask:

> Where, for all its merits, has our mighty intellectual capacity got us? Right to where we are: an unjust world, threatened self-extinction, untold and needless pain and privation in a universe of abundance, the spread of alienation and despair.[23]

These crises are generated by industrial consumer societies. They're the product of lifestyles that are dependent upon an extensive technological base. They are consequences of the way that we live our lives. To some degree, they affect every single one of us. Therefore, various technological interventions for things like obesity, mental illness, and widespread trauma can only ever be palliative. They address the symptoms of the problems rather than the root causes. This also casts doubt on the claim that for every problem there is a high-tech, marketable solution. A radically different approach is demanded.

Climate Breakdown

I shouldn't need to go into too much detail about climate breakdown. A *Guardian* headline, 9 January 2024, says it all: '2023 smashes record for world's hottest year by huge margin'.[24] An exponential increase in the use of fossil fuels like coal and gas since the beginning of the industrial era has resulted in the release of greenhouse gases into the atmosphere. These greenhouse gases trap heat and are significantly increasing the average global temperature with increasingly catastrophic results. As I write, much of the world is experiencing endless heatwaves, wildfires, droughts, and floods. This extreme weather is generating much suffering, threatening already pressured wildlife, and reducing harvest yields. All this is happening significantly sooner than anticipated.

Where will climate breakdown lead? A currently all too likely future is outlined in Bill McGuire's terrifying book *Hothouse Earth*. McGuire lists the unfolding consequences of pumping billions of tonnes of greenhouse gases into the atmosphere over many decades. These consequences include collapsing ice sheets, acidic oceans, a wandering jet stream, storms, floods, killer heatwaves. Human consequences mean starvation, thirst, social and political disruption, mass migration, and climate wars.

McGuire thinks that we're crossing a tipping point. On the far side of this tipping point is Hothouse Earth. Hothouse Earth is a planet where

> lethal heatwaves and temperatures in excess of 50° C (122° F) are nothing to write home about; a world where winters at temperate latitudes have dwindled to almost nothing and baking summers are the norm; a world where oceans have heated beyond the hope of no return and the mercury climbing to 30° C+ (86° F+) within the Arctic circle is no big deal.[25]

This is a planet that is far less habitable than the one that humanity inherited. McGuire writes of the danger of 'wet bulb' temperatures. 'Wet bulb' is a measure of temperature plus humidity. It turns out that a 'wet-bulb' temperature over 35° C means that human bodies can't cool themselves by sweating. This means that a wet-bulb event of 35° C can potentially kill within about 6 hours.[26]

The consequences of climate breakdown for human civilisation are likely to be dire. The ecological philosopher Rupert Read, who was also one of the founders of Extinction Rebellion, writes of likely 'eco-driven social collapse'.[27] He suggests that climate breakdown represents a fundamental failure on the part of humanity, safeguarding the wellbeing of our children. This is because it will 'kill far more in the future than it has done yet'.[28] Our descendants, Read suggests, represent our most 'fundamental care'. We should also have known better. Read calls climate breakdown a 'white swan', something that was predictable from our actions. You can't pour gigatonnes of carbon dioxide into the atmosphere for decades without consequences. He also notes the ubiquity of climate denial. Climate change is still a 'spectre' for many, not seeming quite real.

Read thinks that these large-scale failures of philosophy and practical action will likely lead to one of three possible futures. The first possibility is that society will profoundly transform itself in a sustainable direction (Read calls this 'butterfly'). The second is that our current society will collapse, to be followed by a successor civilisation (dubbed 'phoenix'). The final possibility is that humans and possibly the whole biosphere will become extinct ('dodo'). So climate breakdown challenges industrial consumer civilisation in the most fundamental way imaginable.

Technotopian solutions to climate breakdown tend to focus on technological innovation. This might include things like nuclear fusion, space-based solar power, electric vehicles, lab-grown protein, and geoengineering. I'm not going to argue the merits or otherwise of these sorts of solutions here. But I will say that technofixes seem very unlikely to address the dysfunctional worldviews and mindsets that created the 'polycrisis' in the first place.

Climate activist Charlie Hertzog Young suggests that the extractive, colonial mindset has created a sort of widespread derangement that allows the destruction of the natural world to continue apace.[29] This dimension is typically ignored in technotopian solutions. In fact, the prescription seems to be 'more of the same': untrammelled technological development paired with endlessly extractive capitalism that's now set to expand into outer space. The inflexibility of this approach becomes very clear with one particularly extreme variant of technotopianism. A variant whose adherents claim is essential for the very survival of global civilisation.

Transhumanism: A Technotopian Non-Solution?

Transhumanism is perhaps the ultimate expression of the 'technotopian' future vision. The transhumanist philosopher Nick Bostrom defines it as 'a way of thinking about the future that is based upon the premise that the human species in its current form does not represent the end of our development but rather a comparatively early phase'.[30]

For transhumanists, the future development of humans is not to be left to evolution but to rational, planned engineering. Bostrom writes that 'brain–computer interfaces and neuropharmacology [drugs] could amplify human intelligence, increase emotional well-being [...] and even multiply the richness of possible emotions'.[31] Gene editing can eliminate defects. Pain can likely be eliminated, to be replaced by endless pleasure. Transhumanists also want greatly increased longevity, or even to achieve immortality.

Artificial intelligence is closely linked with these ideas. Transhumanists have written of the 'Merge', which is the allegedly imminent union of humans and machines. One aspect of this project is the development of brain–computer interfaces, which will allow humans to control machines directly with their thoughts. These interfaces also allow a kind of 'mind reading'. Another is the prospect of what has been called 'whole brain emulation'. This is the dream of mimicking the brain entirely within a computer. According to some transhumanists, this might allow us to 'upload' our consciousness onto the cloud. We've seen that this sort of dream is a cornerstone of 'longtermism'.[32]

Bostrom sees transhumanism as an extension of more traditional humanism. He defines it as 'an outgrowth of secular humanism and the Enlightenment'.[33] It's also closely allied with basically liberal values that emphasise the freedom of the individual and the importance of individual autonomy.[34] Transhumanists think that human nature is best improved by 'the use of applied science and other rational methods'.[35] Allied to this is a significant confidence in the power of human beings rationally to choose the best ways to improve the species and also confidence that our technologies will eventually get us where 'we' want to go. The goal is to create posthumans who in multiple ways are superior to present-day humans.

Allied to this is a distrust of nature, something that Rushkoff also noted as a feature of 'the Mindset'. Bostrom writes that

> nature's gifts are sometimes poisoned and should not always be accepted. Cancer, malaria, dementia, aging, starvation, unnecessary suffering, cognitive shortcomings are all among the presents we wisely refuse. Our own species-specified natures are a rich source of much of the thoroughly unrespectable and unacceptable — susceptibility for disease, murder, rape, genocide, cheating, torture, racism.[36]

Perhaps understandably, given this perspective, advocates of transhumanism see their programme as a top priority for humanity. Another transhumanist philosopher, Mark Walker, makes the claim that we ought to create posthumans.[37] He states that 'even though creating posthumans may be a very dangerous social experiment, it is even more dangerous not to attempt it: technological advances mean that there is a high probability that a human-only future will end in extinction'.[38] So for Walker, the transhumanist path is the only way that humanity will be able to move beyond the 'polycrisis' in which we find ourselves.

How might posthumans help? Walker envisages future beings who, through the extensive use of technology, will be our moral, intellectual, and emotional superiors: 'smarter' and 'more virtuous'. The posthumans should be able to 'solve many of the intellectual and practical problems that stymie us'.[39] These superior beings will naturally lead the rest of humanity; 'the best candidates amongst us to lead civilisation through such perilous times are the brightest and most virtuous: posthumans'.[40] Walker even claims that there is a '90% chance of civilization surviving the next two centuries if we follow the transhumanist path', but a less than 20% chance if we don't.[41]

The Devil's in the Detail

Transhumanism's aims might seem superficially desirable. There's certainly too much suffering in the world, and the thought that one day much of it could be eliminated with the flick of a switch is appealing. Increased longevity and health is also appealing, especially to those of us in middle age and beyond. Many teenagers would likely respond enthusiastically to the promise of tech-enhanced 'superpowers'. And the notion that one could live as a virtual self in a virtual world, cut off from any worldly woes seems an extension of the way many of us live now. But as ever, there's a significant gap between theory and practice.

Transhumanism has long been closely allied with eugenics. Eugenics is the idea that social and cultural problems can be solved by breeding better people. The person who coined the term transhuman, Julian Huxley (1887–1975), was also a eugenicist.[42] Eugenics has a significantly tainted history, being closely associated with racism in the United States, where eugenicists encouraged things like forced sterilisation and racial purity. Most notoriously, eugenics is directly linked to the exterminations of the Nazi holocaust.[43]

Despite this, eugenic practices are still promoted today, but they're generally given different labels. The transhumanist philosopher Nick Bostrom was also a founder of longtermism. Longtermism also has partly eugenic underpinnings. Émile P. Torres, an intellectual historian and leading critic of longtermism, has drawn direct parallels to the older eugenics movement. Torres claims that the philosophy retains racist elements and associations, as in a reference to 'dysgenic pressures' in one of the founding documents. ('Dysgenic' was a word used by eugenicists, often in a racial context).[44]

Transhumanism and longtermism also have a significant class dimension. Essentially they're elite projects, which is evident because of their billionaire support. It's understood that not everybody will be able or willing to modify themselves or their children. In today's culture of gross economic inequality, it seems likely that transhumanism will be the preserve of the rich. This is already working out in practice, as billionaires seek immortality and enhancement.

This class dimension means that claims about 'posthumans' potentially leading us to a better world is in one sense deeply conservative. The world is already dominated by billionaires. The advent of 'posthuman' billionaires seems in many ways a logical development of the current economic and political moment. But if this is so, then it seems unlikely that a posthuman elite will raise the chances of humanity's survival to '90%' as philosophers like Mark Walker like to claim. It seems more likely that the existing social order will be preserved with all the structural problems that go along with it. This puts another big question mark over claims that we must develop 'posthumans' in order to secure humanity's long-term survival.

Transhumanism and the Machine-Brain

These are by no means the only difficulties. The implementation of transhumanist projects has already proved to have significant ethical problems. Elon Musk's Neuralink company is one example. Neuralink is currently focused on developing brain chips to help paralysed people, but the ultimate aim is to boost cognition and become 'superhuman'. In 2022, Musk faced animal abuse allegations as an animal welfare group claimed that monkeys had suffered from 'crude surgeries' giving them 'extreme psychological distress'.[45] This is an unfortunate example of the suffering that too often underlies an extractive view of mind and life.

There are other problems with this vision. Transhumanism is rooted in the idea of the brain as a computing or 'information-processing' system or machine. Another basic assumption is that the brain has discrete modules that you can 'read'. If this were true, then it would justify claims that humans and machines are essentially identical and can be 'merged'. But we've already seen that this is a very dated picture. This leads transhumanists to seriously underestimate the difficulties of 'enhancing cognition'. Critic Susan Levin states that 'Across the board they fail to appreciate the complexity of our mental operations.'[46]

Levin:

> Transhumanists' understanding of the brain is […] flawed. Their presumption that particular mental capacities are tethered to specific areas of the brain—and could, therefore, be targeted for manipulation—is increasingly outdated. […] A monumental shift in the focus of neuroscientific research, from discrete areas with dedicated functions to complex functional networks, is well underway. […] Mental tasks such as attention, memory, and creativity engage numerous areas of the brain […].[47]

The brain is coming to be seen as a flowing whole of complex processes and not a modular machine. This renders many of the ambitions of transhumanists redundant. Enhancements will not work, Levin claims: experiments with drugs already show trade-offs between memory, attention, and flexibility.[48] 'Enhance' attention, and you find deficits in memory, and vice versa. So it seems that we're not going to have technology-enhanced mental 'superpowers' any time soon. Which means that the suffering of monkeys and other animals was for nothing. And if the brain and a computer are not, after all, interchangeable, then it seems very unlikely that we'll be able to 'upload' ourselves into an entirely virtual world. Digital immortality looks like a nonstarter.

What's Missing

Douglas Rushkoff has suggested that the transition from human to machine envisaged by technotopian advocates comes at a cost. The companies selling human enhancement of various kinds are 'actively picking which features of humanity to enhance and which to ignore'. Rushkoff concludes that 'The human traits not favoured by the market will surely be abandoned.'[49] In other words, achieving a true, posthuman state might well come at an unacceptable cost.

My suspicion is that a good percentage of exceptional human experiences might well be jettisoned in such a transformation. These things do not really fit into a purely mechanistic world-picture. What lies beyond a mechanistic, materialist, extractive, and ultimately colonial approach cannot be approached in mechanistic, materialist, extractive, and colonial terms — at least, not without destroying what gives these experiences their power in the first place. For me, exceptional human experiences point to a very different way of being in the world and a consideration of them opens up fresh possibilities. This is a perhaps necessary counter to the adoption of 'the Mindset' which acts to shut alternatives down (surely what colonialist projects are bound to do).

The enriched ways of being suggested by exceptional human experiences may also be crucial in the turbulent decades to come. And it's to this possibly that we'll turn in the concluding chapter.

Chapter 10

A Transpersonal Tomorrow

At 8:15 a.m. on Sunday, 5 November 1961, the Van Nuys signal office received notification of a bush fire blazing at the north end of Stone Canyon, on one side of Mulholland Drive.[1] The week was a hot one with high winds and low humidity. The bush fires in Stone Canyon marked the start of the Bel Air fire which burned through 16,090 acres and destroyed 484 homes, damaging an additional 191. The blaze was eventually contained by the efforts of 2500 firefighters and the use of 12 aerial tankers that dropped water on the flames.[2] One of the homes that was burnt was owned by the writer Aldous Huxley and his last wife, Laura. Fortunately they were unharmed, but the fire was otherwise devastating: Huxley lost all his diaries, books, and letters.[3] One manuscript was rescued and had only been a little singed. This was the text of Huxley's new utopian novel *Island*.

Island is in many ways the polar opposite of Huxley's 1932 dystopia, *Brave New World*. *Brave New World* depicts a highly ordered future world of total technological control and mass-produced people. Babies are born in artificial wombs and subsequently 'decanted' for socially predetermined roles. Social control is largely achieved through pleasure, including casual sex and with the use of drugs like *soma*. The novel was intended as a satire on the sorts of futures being proposed by writers like H.G. Wells and the Marxist scientist J.D. Bernal.

Island is different. It's set on the island of Pala in the Philippines. The utopia of Pala shares several elements with *Brave New World*: the inhabitants practise selective breeding, free love, and also use a mind-altering drug, dubbed *moksha-medicine*. However, the kind of society depicted is very different. The basic philosophy is a kind of mix of Buddhism and Vedantic Hinduism. Governance is very light, and people live in decentralised small communities. The spiritual element of society is central. Palans practise meditation and yoga and they use moksha-medicine as a kind of initiatory practice. This 'medicine' allows a glimpse of the higher realities Huxley described in his nonfiction books *The Doors of Perception* and *Heaven and Hell*. One of the founders of the society released a number of mynah birds around the island. The mynah birds have been trained to say 'Here and now, boys! Attention!' This is intended to keep the Palans' minds on the present, a key spiritual practice.

Today, one might smile at the naivety of Huxley's vision. Taking psychedelics does not automatically make you enlightened in the way Huxley seemed to hope. It's possible to be critical of his biases as a white male, especially on topics like Vedantic Hinduism and Mahayana Buddhism. I've also heard several people complain that *Island* is, compared to *Brave New World*, a little dull. But these critiques do not really detract from what Huxley was trying to do. He was trying to depict a humanitarian alternative to the high-tech, over-organised future that he saw nascent within the rapid development of technological mass society. And it's not a coincidence that he saw transpersonal states and psychedelic drugs as a key part of this utopia. So *Brave New World* and *Island* stand for two contrasting ideas about future human societies.

We've already met the twenty-first-century incarnations of *Brave New World*. These sorts of futures are pushed by the advocates of transhumanism and longtermism. They are to be created by exercising ever-increasing amounts of techno-scientific control on nature and on human beings. *Island* can be taken as representative of a different kind of future, which Jennifer Gidley has called 'human centred'. 'Human-centred' futures are based upon the view of human beings as 'kind, fair, consciously evolving, peaceful agents of change with a responsibility to maintain the ecological balance between humans, Earth and Cosmos'.[4]

In this final chapter, I'm going to make the case that the transpersonal and 'paranormal' aspects of human beings fit better within human-centred futures than within the 'technotopian' vision. I'm going to suggest that Huxley was correct to include transpersonal and mystical states as key parts of his utopia. And I'm also going to claim that these 'discarded' elements of human experience might help to improve the quality of people's inner lives in the future as they have in the present and the past. But before this, I'll need to address some potential stumbling blocks.

Anti-Magic: Materialism Requires Loyalty

Scientific materialism commands political loyalty in industrial consumer societies. Support for it can even be seen as synonymous as support for modernity, democracy, and secular society as a whole. Allied to this is a strong sense that anything resembling 'magic' — or worse, 'religion' — should be rejected as anti-science. This is essentially the position of contemporary skepticism, which reflects the tacit views of very many academic scientists. The converse of this is that any rejection of materialism is also seen as a rejection of democracy, science, reason, secular society, and modernity itself. Mystical and paranormal experiences, broadly speaking, seem to challenge scientific materialism. Therefore, according to this mindset, they also challenge modernity. This form of 'guilt by association' remains the basis for a basically political rejection of anything 'mystical' or 'paranormal'.

On the political left, this rejection is reinforced by a different sort of guilt by association. This is because some forms of mysticism and the 'occult' are associated with fascism and Nazism in particular. And it's certainly true that 'occult' beliefs were pervasive in Nazi Germany, although Hitler was apparently not interested in the occult.[5] The historian Nicholas Goodrick-Clarke has argued that Nazism was influenced by a particularly racist form of occultism called 'Ariosophy', which means 'the wisdom of the Aryans'.[6] Heinrich Himmler also incorporated pagan and Grail elements, including the use of runes, into the SS military symbolism.[7]

Today, the association between magic and occultism and far-right politics seems to have been reinforced, especially post-Covid. The progressive journalist Naomi Klein has raised the alarm concerning a new alliance between certain wings of the New Age counterculture with right-wing conspiracy theory culture.[8] The *Conspirituality* podcast runs a skeptical commentary on the alliance between New Age wellness 'gurus' and the online promotion of right-wing conspiracies. The hosts, Derek Beres and Matthew Remski, have also discussed the nature of consciousness on *Conspirituality* and explicitly state their support for the idea that it is entirely generated by the brain.[9]

It's difficult not to see this advocacy of materialism as partly politically motivated, given the association between the New Age, right-wing conspiracy theories, and 'alternative' views of consciousness. This association shows that adopting a post-materialist worldview does not automatically make one virtuous. The podcast is also a good showcase of the dangers of an overreliance on personal revelation, undue influence, and the abandonment of critical thinking.

So Beres and Remski's position is understandable. It's also essentially defensive. We live in an era when the secular space of many countries is under threat from fascism, religious fundamentalism, and other extreme belief systems. This secular space is precious, fragile, and infinitely preferable to the alternatives. At the same time, guilt by association should not determine one's position on consciousness or any other scientific or philosophical topic. Ideally, one's position should be determined by a rational exploration of differing perspectives and of the available facts.

In addition, the idea that an interest in the esoteric is a 'right-wing thing' might be misplaced. Gary Lachman has suggested that occult politics seem actually to have had a broadly progressive character from the 1600s to about World War I. This war was a 'devastating watershed' that provoked, according to Lachman, a 'strong reactionary turn' which he suggests 'isn't surprising, given that a similar *volte face* gripped other manifestations of Western consciousness at the time'.[10] Interest in the occult and paranormal, Lachman suggests, has overall never been politically exclusive.

Perhaps also the political associations of 'magic' and the 'occult' are beside the point. If psi phenomena are 'facts of nature' and extraordinary human experiences are persistent parts of human life, then their acceptance should not depend upon political loyalty. This would be like saying that only left-wingers can accept electromagnetism, or only right-wingers the possibility of extraterrestrial life. What matters from a scientific perspective is a better understanding. What matters from a humanitarian perspective is an exploration of the potential benefits of exceptional human experiences. But here we run into another problem.

The Transpersonal as a Tool

In 1979, Usborne children's books released the *Usborne Book of the Future*.[11] This lavishly illustrated volume depicted life in the not-too-distant future. It predicted many things that have already come to pass: personal communications devices, wall-sized TV screens, electronic mail, and robots. It was often over-optimistic: one spread showed the Olympic Games on the moon in the year 2020. And there was another spread that today might be considered unusual. This was titled 'Mind Over Matter — the Final Frontier'.

These pages discussed the astronaut Ed Mitchell's telepathy experiments on Apollo 14, Yuri Geller's spoon bending, poltergeists, and healing by the 'laying on of hands'. The book suggested that scientists had tried to prove the existence of psi 'without conclusive evidence for and against'.[12] However, if these powers did prove to be real, 'in the distant future people might be doing things that seem like magic to us'.[13] The pages included a fine illustration of a starship crewed by 'ESP sensitives'.[14] The captain steers the ship with the power of his mind, 'machines and weapons are thought-controlled'. In the ship's hospital, psychic healers take care of battle casualties.

Today, it seems more likely that such a ship would be piloted via a brain–machine interface rather than psi. We already have brain–machine interfaces that allow the operation of pieces of technology like drones. But the scenario highlights an important, common approach that's become so ingrained in our culture that it's often unconscious. This is the idea that something like psi must be useful in a practical way for it to have any value or worth. Better still, for something to be really useful, it needs to be integrated into technology.

Think about electricity. Electricity has a very long history. Ancient Egyptian writers knew of electric fish in the Nile, and ancient Mediterranean cultures knew that substances like amber could be rubbed to produce a charge.[15] Lightning was a well-known but mysterious phenomenon. After 1600, electricity was even studied in a more systematic way. However, it didn't start to become 'useful' to human beings until the late eighteenth century. This was when it became possible to store electricity in crude batteries and use it to generate light and heat. The rest, as they say, is history. So electricity was a natural curiosity for millennia before it became instrumentalised. Being *instrumentalised* means being made into an instrument to achieve a goal. Today, of course, electricity powers human society.

The *Usborne Book of the Future* was essentially imagining a time when psi had become harnessed or instrumentalised in a similar way. This has already been tried: the remote viewing programme Project Stargate represents an attempt to instrumentalise psi processes for the purpose of espionage. This attempt has to be seen in the broader context of utilising a range of alternate states of consciousness for practical, and not always ethical, purposes. The remote viewing programme was contemporaneous with the notorious mind-control experiments of MK-Ultra, an illegal programme undertaken by the CIA in America.[16] A more recent example is Silicon Valley workers using 'microdoses' of LSD to service 'productivity' and 'creativity'.[17]

Dean Radin, on the website of the Institute of Noetic Sciences, outlines a number of possible applications of psi.[18] He lists seven potential areas of application: communication and control, forecasting, healing, intelligence, archaeology, dowsing, and counselling. The 'communication and control' section discusses the possibility of creating a psi switch. This is derived from what are called micro-PK (micro-psychokinesis) experiments where participants seem able to very slightly influence the output of a random event generator. Once again, the underlying philosophy does not seem too different from that which created brain–machine interfaces, with the addition of psi as part of the control system. The aim is basically to access or create 'superpowers' via technology, psi or not. Elon Musk, for example, has claimed that his brain-chip company Neuralink aims for a 'Luke Skywalker' solution, a reference to 'the Force' in the *Star Wars* movies.[19]

Instrumentalising psi or mystical experience is essentially a technotopian approach. Douglas Rushkoff points to Silicon Valley operators dropping psychedelics at the Burning Man festivals, with the aim of reformatting their 'cognitive hard drives'.[20] This is a reframing of mystical experience in both instrumental and mechanistic terms. As Rushkoff warns, using magic mushrooms in this way seems unlikely to shift the user's 'basic nature'.[21] This is also very different from the use of moksha-medicine in Huxley's utopia, or from the traditional uses of altered states in Buddhism, Hinduism, and indigenous religions.

The lesson is that the simple presence of psychedelics, psi, or mystical experience in industrial consumer societies hasn't produced a noticeable shift in their economic or cultural trajectories. Instead, altered states have become subsumed into the economic, social, and cultural patterns displayed within industrial consumer societies. (Rushkoff even suggests that the Burning Man psychedelic initiations serve the same purpose that alcohol did for media and advertising executives in the mid-twentieth century.)

This contradicts Huxley's hope, echoed by quite a few members of the counterculture in the 1960s, that psychedelics would trigger an 'inner revolution' for society. Put simply: they didn't. The Age of Aquarius never really dawned. The longer-term result has been more like *Brave New World*.

Reality, Consciousness, and Psi

There's another, potentially more serious, objection to the suggestion that transpersonal, paranormal, and mystical experiences might be significant within a more humanitarian future. This objection states that we can't make claims about the benefits of psi until it's been proved beyond doubt to be real. This sort of objection can be extended to mystical experience, which might, after all, still be due to abnormal brain functioning. So for some, paranormal and transpersonal experiences are essentially 'subjective' and have no basis in 'objective' reality. 'Subjective' things are not real and so can't really benefit anyone.

This objection rests on two assumptions. The first is that there's a clear, obvious difference between objective and subjective reality. There are things that are 'real' like rocks, rainbows, and blood flow in the brain, and there are things that are 'illusory' like dreams, fantasies, and delusions. The skeptical 'smoke without fire' theory would place paranormal phenomena in the second category and explanations involving brain function in the first. From this perspective, in psychedelic trips, the entropy that's measured in the brain is real, objective, and publicly measurable, but the experience of profound unity with a deeply interconnected, living universal consciousness is subjective, private, and doesn't reflect the 'real' world in any significant way.

Henry Bauer, an ex-professor of chemistry and writer on scientific unorthodoxies, claims that this objection is partly cultural. He suggests that we live in an unconsciously 'scientistic' culture, and that when we say that something hasn't been 'proved' scientifically, we mean proved in the sense of the 'hard' sciences like physics, chemistry, and biology.[22] I'd concede that things like 'psi' are not subject to the levels of proof that one finds in the hard sciences. However, I would say that there's cumulatively reasonably strong evidence for psi from the perspective of the social and human sciences. For me, the cumulative evidence provided by transpersonal and paranormal experience, coupled with laboratory psi experiments, points quite strongly in the direction of human beings and other living things having some kind of nonlocal connection with each other and with the world. And I think that mystical experiences in particular suggest that such connections are not impersonal or mechanical but are personal and in some sense animated by a broader, encompassing consciousness.

However, there's a deeper problem here concerning consciousness itself. Exceptional human experiences also challenge the simplistic split between the 'objective' (real) and the 'subjective' ('unreal', 'illusory'.) To understand why, it's necessary to look at concepts like *real, unreal, subjective,* and *objective* a little more deeply.

Most of us, most of the time, are naive realists about the world. We assume that houses, trees, cars, mountains, and people are real and exist in a world that everyone can access through their perceptions. This physical world is 'realer' than the worlds we might encounter in dreaming or in trance states. It is primary, and the worlds of dreams and trances are secondary. True, there might be times when we're not sure whether an experience is real or illusory (conjuring tricks for example, or when people have delusions in severe mental illness), but for the most part, it's easy to distinguish the real from the unreal. This assumption is also mostly tacitly accepted by scientists. Biology, chemistry, and physics are assumed broadly to describe this 'real world': so water is literally made of hydrogen and oxygen; the brain is literally partly composed of neurons and glial cells; evolution describes how living things change over deep time; etc. Most of the time, too, this literalism is taken for granted by both scientists and laypeople.

However, the very mainstream claim that consciousness 'is' a hallucination casts serious doubt on naive realism and even possibly on realism at all. I think it casts doubt on realism in a way that's not yet been fully appreciated by anyone, including idealists, who think the universe 'is' mind. This is because the only access humans have to reality is via the 'hallucination' of consciousness. This means that any theory we have about the 'real' world depends upon the accuracy of the data we're receiving through the 'hallucination'. But there's no guarantee that the world we perceive in the collective hallucination of consciousness accurately maps to any theoretical 'real' world at all.

Some cognitive scientists have understood that this likely undermines realism, at least in the sense that there is a physical world that is 'just' out there. One of the most high-profile scientists to understand this in recent years is Donald Hoffman. In his book *The Case Against Reality*, he suggests that we've not evolved to perceive reality at all, just what helps us to survive and reproduce.[23] Instead, Hoffman suggests that we perceive 'desktop representations' of reality. He calls this approach 'conscious realism'.

I'm going to suggest that even Hoffman does not really appreciate how deeply this undermines traditional ideas of reality. If what researchers like Hoffman suggest is true, and we live in some kind of 'virtual reality', then realism seems fatally undermined. For example, he writes about idealist theories of physics. Yet he still discusses theories of evolution and brain-function in realist terms. But if we really were just experiencing a virtual reality world, then how could we know that these theories were any more 'real' than those of physics? The mathematician and philosopher Alfred North Whitehead saw this problem a century ago:

> Some people express themselves as though bodies, brains and nerves were the only real things in an entirely imaginary world. In other words, they treat bodies on objectivist principles, and the rest of the world on subjectivist principles. This will not do; especially when we remember that it is the experimenter's perception of another person's body which is in question as evidence.[24]

So Hoffman's doubts about reality actually end up being quite selective.

I think that this problem with consciousness means that the idea that there is a simple split between a 'real' objective, publicly observable world and an 'unreal', subjective, private, inner world is fatally undermined. There do, however, exist alternative ways of approaching this problem which I think are very relevant for paranormal and mystical experiences. In a discussion with scientists, the Dalai Lama described a traditional Buddhist classification of phenomena that does not split the world into objective or subjective at all. Instead, the split is between 1) *evident phenomena,* known through direct perception, 2) *remote or obscure phenomena,* known through inference, and 3) *extremely remote or obscure phenomena,* which are known only through the testimony of a third person.[25]

Science as we understand it only deals with *evident phenomena*. Spiritual traditions do address *remote or obscure phenomena* by claiming that spiritual practices can allow access to states of consciousness where someone can verify the claims of others. Because *extremely remote phenomena* are third party, only they cannot be verified by science or by spiritual practice. However, they still count as kinds of experience, just not consensual ones.

This three-way split seems also to me to be a more satisfactory way of approaching exceptional human experiences as a whole without the stumbling block of categorising them as either strictly 'objective' or 'subjective'. Psi in particular seems to cross the boundaries of all three categories; it is at once both 'objective' and 'subjective'. To put it another way, it seems to leave traces in the public world but also reside within personal consciousness.

The assertion that consciousness 'is' a hallucination also has implications for society. Paul Devereux has suggested that if we really attend to what cognitive scientists are suggesting, then we have to acknowledge that culture, including the scientific aspects of culture, are also hallucinations.

> We only think the consensus Western worldview is "reality" (which is—what?—a special or strong form of hallucination?) because some of the ways it arrives in the brain, though not how it arises in the mind, can be measured by the means we use and accept. But the Western worldview is highly dependent on its cultural environment, and extrasensory, transpersonal experiences are equally accepted as part of reality in

> other worldviews, and can become so even to Westerners who adapt to those worldviews [...].[26]

This strongly suggests that a deeper understanding of the significance of extrasensory, transpersonal experiences will not be possible for those who are deeply immersed within the hard-line materialism of industrial consumer cultures. Psi experience will, as we've seen, get pushed to the margins and remain unacknowledged by the major institutions of culture, which will be mostly unconscious of them. This will include, for the most part, neuroscience and biology.

What this means is that neuroscientists and consciousness researchers who believe that humans are nothing but 'beast machines' have created a hermetically sealed world that excludes anything that doesn't match up with their assumptions. This world is self-completing, mostly airtight, and is essentially determined by attention. Reality is that which you 'tune into'.[27] That which is 'tuned out', in this case the extrasensory and transpersonal experiences of which Devereux spoke, is by definition non-existent. This is why researchers and skeptics can state with confidence that there's 'no' evidence for things like psi. Any potential evidence is simply tuned out of their consciousness on a personal and cultural level. But what if, instead, the possibility of psi and the transpersonal were more generally 'tuned in'?

An Ancient Faculty?

The biologist Sir Alister Hardy believed spiritual awareness to be a human basic. He believed that it naturally developed in the course of evolution and that it was as integral to human beings as language. However, his attitude towards possible spiritual realities was not materialistic. He openly discussed the possibility of telepathy in animals in the 1960s and also participated in a mass psi experiment.[28] In the 1990s and 2000s, a protégé of Hardy's, David Hay, suggested that the evidence strongly supported Hardy's theory that 'spiritual awareness is a human universal, part of our biological makeup that has evolved through the process of natural selection because it has survival value'.[29]

If this is true, then it's likely that such a faculty is ancient. David Lewis-Williams provides evidence for this in *The Mind in the Cave*, but there's plenty of independent archaeological evidence for a spiritual or at least ritual sense in prehistory. This has implications for the future. If the faculty is ancient, then it likely has longevity. For statistical reasons, if something has already lasted for, say 10,000 years, then there's a good chance that it will still exist in 10,000 years' time. The principle is that things that have already been around for ages are likely to be around for ages yet.[30]

If Hardy's theory is right, then this means that even the most ferocious campaigns of rationalists, skeptics, and atheists are unlikely to be able to eliminate a basically spiritual sense. The campaigns are likely to be about as successful as puritanical campaigns against sex. Conclusion: this side of human experience is likely to persist, no matter what. So how could it be incorporated into a more humane future? And is something like Huxley's utopia really at all feasible in reality?

David Hay was in no doubt that the suppression of this spiritual awareness, common in industrial consumer societies, is damaging. He believed that

> many of our most pressing social and political problems—meaninglessness, the collapse of a sense of human community, the draining away of trust and social capital in general, the turning of everything into a commodity, and carelessness about the ecology of the planet—have their origin in the ignoring of the aspect of

> our human nature adapted to deal with them, relational consciousness or spirituality.[31]

What would individuals and society have to gain by re-embracing relational or spiritual modes of consciousness? We already have some very significant clues. One example is healing. The physician Gabor Maté reported his experience on an ayahuasca retreat in Peru with Shipibo shamans.[32] Maté has carried trauma all his life, from the time he was a baby and had to be rescued from Nazi persecution by being sent away by his parents. Maté reported that on the retreat, after some difficulty, he had a profound and visionary psychedelic experience that helped him to heal at least a portion of that trauma. He described being 'thrown down with sudden, involuntary force' on his mat and experienced a trance for about 2 hours. During this time he had a vision of a deep blue 'sky screen' on which were written in 'giant cloudlike wisps of letters' the word BOLDOG, which in Hungarian means 'happy'.[33]

Despite this experience, Maté claims not to be a psychedelic 'evangelist'. He does not believe that psychedelics on their own will transform either healthcare or larger human consciousness. This will, Maté believes, need 'vast scale social change'.[34] Still, the healing practice of the Shipibo shamans shows that the experience of profound and visionary states, within a supportive cultural context, can be very powerful and personally significant.

Beyond this is the possibility that so-called 'self-transcendent experiences' (experiences where a sense of self is reduced and feelings of connection to other people, nature, and the world are increased) might help to transform mindsets and behaviours towards social and environmental problems. A 2022 paper has discussed this issue, looking at psychedelic and other types of self-transcendent experience. The authors concluded that 'the careful administration of psychedelics in legal settings might hold the potential to unlock a deeper understanding of human beings that goes beyond our material existence and connects us with others and nature'. They also concluded that self-transcendent experiences 'have the potential to act as a catalyst for new ways of being in the world that support ecological wellbeing'.[35] This is pretty much the role of moksha-medicine in Huxley's fictional utopia. The thought that at least some of Huxley's vision might be realised in practice remains a tantalising possibility.

The problem is that we currently live in a culture where relational consciousness is in abeyance. This has consequences for society at large, and science in particular.

Science without Relational Consciousness

Industrial consumer societies, supercharged by hyper-capitalism and enabled by fossil fuel-based technology, have proved highly destructive. Such destructiveness forces a re-evaluation of the metaphysics underlying the philosophy of such a society. It means that claims about 'enlightenment' and 'progress' cannot go unchallenged. We especially have to ask questions about the role and function of institutional science.

In particular, the association of science with scientific materialism needs to be challenged. This is potentially a tall order. We've seen that for many working scientists, especially in biology and neuroscience, science and materialism are synonymous. Moreover, it's sometimes claimed that the success of science is due to it being materialistic. The biologist and skeptic Jerry Coyne labours this point by insisting that the lesson of all science is that materialism is true.[36] This is partly because science is still presented as the polar opposite of belief systems that include divine intervention and magical forces.

Despite this, I'd claim that the success of science is due to its attention to reality and willingness to experiment as opposed to any specific allegiance with materialism. We've seen that it's possible to devise practical experiments that at the very least strain the essentially nineteenth century materialism to which many in biology and neuroscience remain wedded. It's less clear that psi violates twenty-first-century physics, although in its current state physics still doesn't accommodate the deeper problems of consciousness.

Very often, too, progress in science has only happened because basic background assumptions were challenged. For example, Einstein's ideas about quantum theory and relativity were likely influenced by the critics of mechanistic and materialistic views within classical physics.[37] In a similar way, Rupert Sheldrake is likely right to claim that science might be 'set free' if the basic assumptions of scientific materialism are questioned.[38] So science and materialism are not necessarily synonymous.

This questioning is currently urgent. This is because of evident 'cracks in the modern scientific dream', as Allan Kellehear put it.[39] These 'cracks' could perhaps be dismissed if we'd managed to create a more utopian techno-scientific world by the 2020s. The sort of utopian world depicted in the *Usborne Book of the Future*, where a sophisticated technological culture has learnt to co-exist with the natural world, where fossil fuel dependence has ended, where populations are tended by robots, and where outer space is explored peacefully.

Such an idyllic future has great appeal, but unfortunately, it's not where we're headed. At the present moment, we're poised between Rupert Read's three futures of a transformed 'butterfly' society, a successor 'phoenix' civilisation that rises from the ashes of our own, or the 'dodo' of total extinction. The role of institutional science in contributing to the crises that have led us to this point needs to be addressed.

It's true that science has a likely crucial role in the potential transformation of society towards sustainability. The exploratory role of science in discovery of course remains central. Modern medicine has brought countless benefits. Also, science is needed for ecological repair, for humanitarian purposes, and for any transition to a post-fossil fuel society that does not involve total collapse. But what kind of science should that be? This question is especially relevant with respect to consciousness.

Anil Seth and others have claimed that the only science of consciousness that is possible is one where 'metaphysics' are avoided, and where we stick to what is testable only.[40] 'Metaphysics' is in fact treated as if it's a dirty word. This is essentially a pragmatic approach to consciousness, but I've already explored some reasons why it's likely to be inadequate. The main problem is that metaphysics can't be avoided. In practice, 'avoiding metaphysics' is really shorthand for approaching the mystery of consciousness with unchallenged mechanistic and materialistic assumptions. In other words, business as usual. But business as usual is no longer working.

Those same mechanistic, materialistic kinds of science force us to approach the world through a particular, narrow lens. This lens was adopted from the seventeenth-century philosopher René Descartes, who conceived of the world as a giant mechanical machine that lacked sensation or consciousness. Descartes thought that you could understand any entity by breaking it into pieces and studying how the components worked.[41] Once you understand how the pieces work, you also understand how the whole works. This approach became the basis for experimental science and remains a fundamental assumption. It's a foundational practice in biology, neuroscience, and cognitive science, and is known as *reductionism*. The materialistic approach to consciousness works in the same way. It's assumed that once the underlying component 'mechanisms' of consciousness are understood, then consciousness will be fully explained. This is the essence of the 'brain-wiring/information-processing' approach to consciousness.

This particular way of approaching the world has in the past been very useful but has always generated significant problems. The reductionist 'lens' shows a world where relational consciousness has been stripped away. What is left are inert parts in motion. One negative consequence is that the world comes to be seen as a heap of usable resources. This has had problems environmentally because forests, mountains, animals, and people are also seen as resources.[42] Another problem is that people and animals are seen as wholly mechanical, literally 'beast-machines'. However, this is only how people and animals appear through the reductionistic lens. This lens does not reveal the ultimate truth but allows a limited, although in some ways powerful, mode of understanding.

This particular mode of understanding might even have neurological origins. Psychiatrist and neuroscience researcher Iain McGilchrist has suggested that it is essentially how the left hemisphere of the brain sees the world.[43] The mode of attention engendered by the left hemisphere sees the world as dead, manipulable objects. By contrast, the mode of attention enabled by the right hemisphere is more like relational consciousness. McGilchrist believes that we now live in a society that is dominated by left-brain consciousness. The life sciences in particular remain largely sciences of the left brain — McGilchrist calls this an example of 'left hemisphere capture'. This 'capture' produces an unbalanced, biased way of seeing living things, consciousness, and the world.

Former Professor of Sustainability Leadership Jem Bendell has suggested that the current moment of ecological, social, and cultural crisis forces us to recognise the limitations of reductionistic and mechanistic approaches. He writes that he finds the 'natural scientist much less interesting and wise on metaphysical matters than the teachers of the great wisdom traditions'.[44] One reason is that currently many scientists are proving unreliable witnesses: Bendell claims that quite a few have retreated into 'extremist techno-idolatry' in the face of unfolding planetary disaster.

Bendell enlarges his critique through an exploration of free will. He suggests that a purely scientific approach is unlikely to furnish us with final questions on this. One problem is that such theories can only ever describe 'mechanistic influences on a living being's choices', which is why many neuroscientists and biologists have concluded that free will does not exist.[45] However, such conclusions are arrived at by setting up experiments that effectively exclude any non-mechanical factor, like intentionality, from intruding.[46] In other words, neuroscience frames experiments with mechanistic assumptions and gets mechanistic results. This is a kind of self-fulfilling prophecy and an example of the 'capture' of which McGilchist spoke.

Bendell himself takes a position on free will technically known as metaphysical libertarian. This means that he holds that some aspects of humans are free of various influences on them. As Bendell also suggests, this invites questions of a metaphysical nature. How could some aspects of human minds and consciousness be free of mechanical causation?

This thought leads to post-materialist views of consciousness where events in our brains do not have a wholly physical explanation. This also invites a consideration of notions of mind and consciousness found in Buddhist and Hindu texts. Bendell states, 'I believe the matter of free will and of freedom are relevant to contemporary environmental and political philosophy as societies are destabilised [...] relative free will can be regarded as essential within nature.'[47] In particular, Bendell sees the assertion of free will as an essential counter to authoritarian regimes and also as a counter to claims that the human polycrisis is an 'inevitable' consequence of human nature.

So science as it currently exists is necessary but probably not sufficient to tackle the problems with which we are faced.

This seems especially true for questions of meaning, agency, and purpose — questions that concern the 'inner dimensions' of existence. In addition, scientific materialism, which is proving inadequate for current challenges, needs in some way to be transcended. We desperately need broader, more encompassing perspectives. Perspectives that include or are informed by relational consciousness (which McGilchrist thinks furnishes us with a more comprehensive view of reality). This shift is necessary because of the large-scale failures of philosophy and action highlighted by Rupert Read and others.

A Benign Tomorrow?

Consider Rupert Read's alternative futures of butterfly, phoenix, or dodo — standing for a transformed society, a successor society, or extinction. I have little to say about potential extinction, except the following. Humanity has serious flaws, but also qualities and potential. We're likely unique in the universe, even if our technological civilisation is not, ultimately, alone. Extinction would be a tragedy and should be prevented if possible.

Think instead about 'butterfly' or 'phoenix'. Read's descriptions of transformed or successor civilisations are fairly vague. I can imagine dystopian versions of both. For example, a transformed civilisation might be an autocratic one, where people live lives of abject misery. One can picture a totalitarian surveillance state with AI-powered rulers spying on people's every thought.[48] The harsh ecological conditions of the future might make such societies of total control all too easy to justify. Some kinds of 'phoenix' civilisation might also be miserable places to live — think of a feudalism reborn, where warlords fight over patches of farmable land or drinkable water, dominating remnant populations on the edge of starvation.

So quality of life would matter a great deal in either the 'butterfly or 'phoenix' scenarios. My own wish is for a future that's as benign as possible, for as many people as possible, given what are likely to be very tough environmental conditions. I'm reasonably undogmatic about the specific pathway towards this goal, although obviously avoiding dystopia or extinction remains desirable.

I'm going to finish with a bold claim. That the healthy recovery, access, and expression of primary spiritual awareness might be essential in order to realise a benign tomorrow. This might allow the development of what Bendell calls an 'Evotopia', where humanity 'better beholds natural reality' and reforms its destructive relationship with the planet.[49] This sort of shift might only be possible locally, or it might become more widespread. But it seems to me something to aspire to and potentially preferable to the alternatives.

In 2018, I tried to imagine just such a future:

> with sustainable cities [...] surrounded by patchworks of small farms giving way to an extended wilderness. In the forests, big mammals have made a comeback, perhaps assisted by genetic technology. In the human realm, technology betters lives and does not manipulate, dominate, and control. Profit is no longer the predominant driving force [...]. Humans [are] perhaps a little less selfish and a little more generous. They might be less prone to the mental distress and chronic diseases that plague early twenty-first-century life. They'd prioritise happiness, community, and the natural world over material goods.[50]

All this might be facilitated by a re-discovery of relational consciousness and a cultural acceptance of humanity's deep, unexpected links to the cosmos. A broader engagement with psi and mystical experience might be a key part of such a cultural shift.

In this context, I'd suggest that the current programmes of 'spiritual' marginalisation and suppression advocated by skeptics and other well-meaning champions of a secular techno-future have failed. That if the spiritual or transpersonal side of human beings is not accessed in healthy ways, it will tend to find expression in unhealthy ways. Here again, sexuality seems a good analogy.

Aldous Huxley himself understood this very well. In an essay titled 'Beliefs' he suggested that the philosophy of meaninglessness that flourished in the 1920s and 1930s gave rise to things like Nazism and Maoism.[51] These functioned as substitutes for religion and spirituality in a spiritual vacuum. Today's intensely polarised world, with resurgent fascism and fundamentalisms of all varieties, is also arguably in part a consequence of such suppression.

For me, psi, mystical encounters, and extraordinary human experience at least potentially represent a way out of this crisis of meaninglessness. A consideration of these things opens up possibilities rather than closes them down. The totality of the evidence suggests that we live in a world with deep, unexpected interconnections; that the visible world as a whole is far more than mechanical, mindless 'fields and forces'; and that when we smile at nature, just occasionally, nature smiles back.

Endnotes

Chapter 1

1. Internjack, 2015.
2. Conefrey, 2012.
3. Reincarnate Lama Returns to Thame!, 2015.
4. White, 1999.
5. Casad & Luebering, 2023.
6. Matlock, 2021.
7. Attributed to Nietzsche, *Twilight of the Idols*. Quoted in Wilson, 1986, frontispiece.
8. Broad, 1953.
9. Murphy, 1992, p. 600. For a full discussion of these 'limiting principles' see Broad, 1953.
10. Sheldrake, 2020.
11. Riskin, 2016.
12. Gardner, 1985.
13. Wyatt, 2017.
14. Wyatt, 2017.
15. Leal, 2008.
16. NeuroAI UW Laboratory, 2013.
17. Dennett, 1995, p. 203.
18. Dennett, 1995, p. 370.
19. TED2017, 2017.
20. TED2017, 2017.
21. Seth, 2021.
22. Luke, 2019.
23. Prince, 1930.
24. Breaking Convention, 2017.
25. Blackmore, 2020.
26. Seth, 2021.
27. Rivas, Driven, & Smit, 2016.

28. White, 2014, p. 72.
29. White, 2014. Paraphrase of White's ideas by interviewer Linda Heuman, p. 72.
30. Walach, 2019.
31. Weinberg, 1977, p. 144.
32. Weinberg, 1977, p. 155.
33. Walach, 2019, p. 12.
34. Walach, 2019, p. 16.
35. Rushkoff, 2022, p. 61.
36. Martin, 2023, p. 22.
37. Lee, 2021.
38. World Economic Forum, 2023.
39. Harari, 2020.
40. Winter et al., 2023.
41. Mayer, 2008.

Chapter 2

1. Lenharo, 2023a, p. 14.
2. Whitehead, 2004.
3. Horgan, 2016.
4. Whitehead, 2004, p. 69.
5. Whitehead, 2004, p. 69.
6. Whitehead, 2004, p. 70.
7. Blackmore & Troscianko, 2018.
8. Whitehead, 2004.
9. Whitehead, 2004, p. 69.
10. Whitehead, 2004, p. 86.
11. The White House, 2013.
12. Whitehead, 2004, p. 73.
13. Chalmers, 1995, p. 201.
14. Seth & Bayne, 2022.
15. This formulation is taken from Westphal, 2016, p. 1.
16. Descartes, 1997.

17. Downing, 2021.
18. Seth & Beyne, 2022. This claim is very similar to what used to be called identity theory, which is that conscious experiences and brain function are somehow identical. Identity theory had theoretical problems, which led to its abandonment.
19. Searle, 1997.
20. Tallis, 2011, p. 87.
21. Blackmore & Troscianko, 2018.
22. Seth, 2021.
23. Blackmore & Troscianko, 2018, Chapter 3.
24. Frankish, 2017.
25. Schurger & Graziano, 2022.
26. Schurger & Graziano, 2022, p. 4.
27. Strawson, 2018.
28. Seth & Bayne, 2022.
29. Miller, 1956.
30. Baars, 2017; Blackmore & Troscianko, Chapter 5.
31. Tonini, 2017a.
32. Baars, 2017.
33. Discussed in Kelly, 2022.
34. Nemirovsky et al., 2023.
35. Baars, 2017, p. 231.
36. Kelly, 2022, pp. 229–30.
37. Westphal, 2016, p. 127.
38. IIT concerned et al., 2023.
39. Quoted in Lenharo, 2023b.
40. Westphal, 2016, p. 127.
41. Bor, 2012, p. 3.
42. Tang et al., 2023.
43. Cuthbertson, 2023.
44. Rainey et al., 2020.
45. Rainey et al., 2020.
46. Rainey et al., 2020.

47. Kastrup, 2019.
48. These assumptions are described in Westlin et al., 2023.
49. Westlin et al., 2023, p. 252.
50. Westlin et al., 2019.
51. Nahm, Rousseau, & Greyson, 2017.
52. Nahm, Rousseau, & Greyson, 2017, p. 967.
53. Kelly et al., 2007; Kelly, 2015a.
54. Broad,1953.

Chapter 3

1. Quotes drawn from the online edition of Glanvill, 1681.
2. Hunter, 2020.
3. Hunter, 2020.
4. Hunter, 2020.
5. Hunter, 2020.
6. Hunter, 2020, p. 87.
7. Sommer, 2019.
8. Robbins, 1959.
9. Robbins, 1959.
10. Evans, 1982.
11. Summary of Roy Porter's ideas in Introduction to Ankarloo & Clark, 1999.
12. Quoted in Evans, 1982, p. 49.
13. Hunter, 2020.
14. Hunter, 2020.
15. Hunter, 2020.
16. Royal Society, 2015.
17. Royal Society, 2015.
18. Hunter, 2020.
19. Hunter, 2020; White, 1998, discusses Newton's alchemical writings.
20. Hunter, 2020, p. vii.
21. Hunter, 2020, p. vi.
22. Hunter, 2020, p. 1.

23. Sommer, 2019.
24. Sommer, 2019.
25. Sommer, 2019.
26. Wikipedia Contributors, 2019d.
27. Bristow, 2023.
28. Toulmin, 1992.
29. Wikipedia Contributors, 2020c.
30. Josephson-Storm, 2017.
31. Josephson-Storm, 2017, pp. 58–9.
32. Mobbs & Watt, 2011.
33. Sommer, 2016.
34. Josephson-Storm, 2017.
35. Josephson-Storm, 2017, p. 34.
36. Josephson-Storm, 2017, p. 32.
37. Gauld, 1968.
38. Quote in Gauld, 1968, p. 44.
39. Gauld, 1968.
40. Gauld, 1968.
41. Myers 1903/2001, p. 3.
42. Myers 1903/2001, p. 3.
43. Quoted on the inside cover of every journal of the SPR.
44. Gauld, 1968, p. 164.
45. Quoted in Inglis, 1884, p. 40.
46. Cardeña, 2015.
47. Cardeña, 2015.
48. Noakes, 2019.
49. Edwards, 1999.
50. Watson, 1913.

Chapter 4

1. Blair & Blair, 1988. This expedition was filmed for Episode 4 of their documentary, 'Dream Wanderers of Borneo'. (Ring of Fire: An Indonesian Odyssey, 1988).
2. Blair & Blair, 1988, p. 243.

3. Carpenter, 2012.
4. Narby, 1998, p. 4.
5. Narby, 1998, p. 1.
6. Sheldrake, 2020.
7. Josephson-Storm, 2017.
8. Castro, Burrows, & Wooffitt, 2014.
9. Roe & Linnett, 2017.
10. Watt et al., 2014.
11. Wikipedia Contributors, 2019b.
12. Wikipedia Contributors, 2020a.
13. Cusick, 2016.
14. Wikipedia Contributors, 2019a.
15. Lorraine, 2018.
16. Westrum, 2011, p. 10.
17. Westrum, 2011, p. 12.
18. Mosley, 2023.
19. Moss, 1977, pp. 89–90.
20. Moss, 1977, p. 91.
21. Linszen et al., 2022.
22. Gurney, Myers, Podmore, 1886, p. 389.
23. Gurney, Myers, Podmore, 1886, p. 393.
24. Gurney, Myers, Podmore, 1886, pp. 399–400.
25. Gurney, Myers, Podmore, 1886, p. 424.
26. Roe, 2019, p. 5.
27. Quoted in Quaglia, 2022.
28. Radin, 1997.

Chapter 5

1. Duggan, 2019.
2. Eysenck & Sargent, 1993.
3. The online Omni calculator allows you to experiment with coin flips and probability. https://www.omnicalculator.com/statistics/coin-flip-probability

4. The paper by Simmonds-Moore & Holt, 2007, reported chance level results overall, but one by Roe, Holt, & Simmonds, 2003, was significant. The overall positive significant hit rate (32.1%) of all combined Northampton Ganzfeld experiments up to 2024 was reported by Roe, 2024.
5. Blackmore, 1987; Sargent, 1987.
6. Honorton, 1993.
7. Honorton, 1993.
8. Hyman, 1985; Honorton, 1985.
9. Radin, 1997.
10. Hyman, 1985.
11. Hyman & Honorton, 1986. Quoted in Parker, 2017.
12. Bem & Honorton, 1994.
13. Milton & Wiseman, 1999.
14. Parker, 2017.
15. Parker, 2017.
16. Storm, Tressoldi, & Risio, 2010.
17. Williams, 2013, p. 654. Quoted in Parker, 2017.
18. Wikipedia Contributors, 2019e.
19. Reproduced from the original CIA files reprinted in Marwaha, 2020b, p. 11.
20. Marwaha, 2020a.
21. Targ & Puthoff, 1977/2005.
22. Targ & Puthoff, 1977/2005, p. 31.
23. Marwaha, 2020a.
24. Marwaha, 2020a, p. 6.
25. Reproduced in Marwaha, 2020b, p. 10.
26. Marwaha, 2020b, p. 10.
27. Outline of this phase of Stargate's history taken from Marwaha, 2020a.
28. Marwaha, 2020a.
29. Marwaha, 2020a.
30. May, 1996.

31. May, 1996, p. 106.
32. Utts, 1995. Quoted in Schwartz, 2017.
33. Hyman, 1996, p. 31.
34. Hyman ,1996, p. 32.
35. Wiseman & Milton, 1998.
36. Marwaha, 2020a.
37. Tressoldi & Katz, 2023, p. 467.
38. Tressoldi, 2011.
39. Cardeña, 2018.
40. Quoted in Cardeña, Palmer, & Marcusson-Calvertz, 2015.
41. Roe, 2018, p. 7.
42. Rabeyron, 2020.
43. Rosenthal & Jacobson, 1968.
44. Roe, 2018, p. 7.

Chapter 6

1. Blackmore, 1996, p. 8.
2. Described in Blackmore, 1982.
3. Blackmore, 1996.
4. Blackmore, 1982.
5. Blackmore, 1996, p. 242.
6. Blackmore, 2010.
7. Blackmore, 1987a.
8. Sargent, 1987.
9. Roe, 2021, p. 5.
10. Berger, 1989; Wehrstein & Duggan, 2019.
11. Out-of-body experiences, for example, might tell us new things about the brain or consciousness. Blackmore, 2020.
12. See the website https://skepticalaboutskeptics.org/ for specific case studies.
13. Humphrey, 1996.
14. Weiler, 2020.
15. Gieryn, 1999.

16. Bauer, 2001.
17. Bauer, 2001.
18. Martin, 2021.
19. Geller's story is summarised in Playfair, 2015.
20. Playfair, 2015.
21. Puharich, 1974.
22. Green, 2018.
23. Targ & Puthoff, 1974.
24. Targ & Puthoff, 1977/2005, p. 137.
25. Green, 2018.
26. Quoted in Green, 2018, p. 21
27. Quoted in Green, 2018, p. 21.
28. Green, 2018.
29. Sagan, 1996, p. 28.
30. Wehrstein & McLuhan, 2020.
31. Quoted in Wehrstein & McLuhan, 2020.
32. Hansen, 1992.
33. Hansen, 1992.
34. Horowitz, 2020.
35. Wikipedia Contributors, 2020b.
36. Gerbic, 2013a.
37. Gerbic, 2013a.
38. Gerbic, 2013a.
39. Gerbic, 2013a.
40. Gerbic, 2013b.
41. Martin, 2018, p. 379.
42. Martin, 2021.
43. In Krippner & Friedman, 2010, p. 155.
44. Roe, 2017, p. 152.
45. Reber & Alcock, 2020, p. 1.
46. French & Stone, 2014, p. 18.
47. Zingrone, n.d.
48. Kellehear, 1996, p. 265.

Chapter 7

1. Chaikin, 1994/2019.
2. Institute of Noetic Sciences, 2016.
3. Mitchell, 1996.
4. Translations in Hartranft, 2003.
5. Tart, 1990.
6. Luke, 2019.
7. Luke, 2019.
8. Narby, 1999.
9. Luke, 2019, pp. 102–3.
10. Lewis-Williams, 2002.
11. Lewis-Williams, 2002.
12. Lewis-Williams, 2002.
13. Marshall, 2015.
14. Henshilwood, 2018.
15. Interested readers can consult Appendix I of Hancock (2005), which gives a summary of the controversies.
16. Hancock, 2005, p. 348.
17. Hancock, 2005.
18. Marshall, 2015.
19. Professor Ed Kelly has suggested to me that there isn't really a singular type of trance state. It seems more likely that there are families of trance states. See Kelly & Locke, 2009.
20. Huxley, 2009a.
21. Luke, 2022.
22. Luke, 2022, p. 28.
23. Luke, 2022, p. 29.
24. Listed in Luke, 2019, p. 95. From Shanon, 2002.
25. Seth, 2021.
26. Sacks, 2012.
27. Timmerman et al., 2019.
28. Luke, 2019, p. 97.

29. Wikipedia Contributors, 2023.
30. Guerra-Doce et al., 2023.
31. Guerra-Doce et al., 2023.
32. Quoted in Reed, 2008.
33. Reed, 2008.
34. Huxley, 2009a.
35. Reed, 2008.
36. Presti, 2017.
37. Griffiths et al., 2006, p. 284.
38. Grosso, 2015.
39. Grosso, 2015.
40. Carhart-Harris et al., 2012.
41. Carhart-Harris et al., 2012.
42. Keshmiri, 2020.
43. Timmerman, et al., 2022.
44. Timmerman, et al., 2022.
45. Pollan, 2018.
46. Pollan, 2018, p. 389.
47. Kelly & Kastrup, 2018.
48. Seth et al., 2018.
49. Seth et al., 2018.
50. Kastrup, 2018.
51. Kastrup, 2018.
52. Sawyer, 2022.
53. James, 1917/1982, p. 14.
54. Petersen, 2019.
55. Sawyer, 2022.
56. Marshall, 2022. Quote from Ritchie, 2021, p. 285.

Chapter 8

1. Quoted in Carr, 2002.
2. Carr, 2002.
3. Blackmore, 1996.
4. Blackmore & Troscianko, 2018, p. 388.

5. Blackmore & Troscianko, 2018, p. 388.
6. Alcock, 2003.
7. Alcock, 2003, p. 29.
8. Alcock, 2003, p. 29.
9. Watt & Tierney, 2014.
10. French & Wilson, 2007.
11. Bentall, 2014.
12. Blackmore & Troscianko, 2018.
13. Blackmore, 2020.
14. Willin, 2019, pp. 66–67.
15. Playfair, 2011, p. 39.
16. Gregory, 1983. A comprehensive and balanced overview of the Enfield case is given in the 2023 Apple TV series, 'The Enfield Poltergeist'. This features actors lip-synching to audio tapes made during the case. It includes interviews with critics as well as advocates and witnesses.
17. Loftus, 1996.
18. Hall, McFeathers, & Loftus, 1987.
19. French, 2003.
20. Otgaar et al., 2019.
21. McLuhan, 2010.
22. Reber & Alcock, 2020.
23. Albert, 2023.
24. Stapp, 2009.
25. Mroczkowski & Malozemoff, 2019.
26. Radin, 2019.
27. Acunzo, 2023, p. 537.
28. Acunzo, 2023, p. 537.
29. Pilkington, 2013.
30. Quoted in Pilkington, 2013.
31. Marwaha, 2020a, p. 5.
32. May, Utts, & Spottiswoode, 1995.
33. May & Marwaha, 2014.
34. May & Marwaha, 2014.

35. Vernon, 2021.
36. Carpenter, 2012.
37. Carpenter, 2012.
38. Carpenter, 2012, p. 14.
39. Carpenter, 2012, p. 17.
40. Carpenter, 2012, p. 17.
41. Carpenter, 2012, p. 18.
42. Carpenter, 2012, p. 18.
43. Kelly, 2015b.
44. Esalen website, https://www.esalen.org/
45. Kelly, 2015b, p. 3.
46. Kelly, 2015b, p. 4.
47. Kelly, 2015b, p. 4.
48. Kelly et al., 2007.
49. Kelly, Crabtree, & Marshall, 2015; Kelly & Marshall, 2021.
50. See Introduction to Kelly et al., 2007; Kelly 2015a.
51. Petersen, 2019.
52. Underhill, 2002, p. 27.
53. James, 1917/1982.
54. Kelly et al., 2007, p. xxviii.
55. Matlock, 2021.
56. Myers, 1961/2001.

Chapter 9

1. Harris, 2022.
2. Harris, 2022.
3. Harris, 2022.
4. Harris, 2022.
5. Harris, 2022.
6. Harris, 2022.
7. Turnbull, 1997; Devereux, 1996.
8. Midgley, 2014.
9. Gidley, 2017, p. 100.
10. Gidley, 2017, p. 100.

11. Rushkoff, 2022.
12. Rushkoff, 2022.
13. Rushkoff, 2022, p. 58.
14. Rushkoff, 2019, p. 79.
15. Nosta, 2023.
16. Nosta, 2023.
17. MacAskill, 2023.
18. Torres, 2021.
19. In line with Sheldrake's ten 'dogmas' of materialism. Sheldrake, 2020.
20. Maté & Maté, 2022.
21. Maté & Maté, 2022.
22. Taylor, 2023.
23. Maté & Maté, 2022.
24. Carrington, 2024.
25. McGuire, 2022, p. xv.
26. McGuire, 2022.
27. Read, 2022, p. 3.
28. Read, 2022, p. 26.
29. Young, 2023.
30. Bostrom, 2003, p. 4.
31. Bostrom, 2003, p. 5.
32. MacAskill, 2023.
33. Bostrom, 2011, p. 55.
34. Bostrom, 2003.
35. Bostrom, 2011, p. 55.
36. Bostrom, 2011, p. 57.
37. Walker, 2011.
38. Walker, 2011.
39. Walker, 2011, p. 97.
40. Walker, 2011, p. 107.
41. Walker, 2011, p. 108.
42. Rifkin, 1999.
43. Rifkin, 1999.

44. Torres, 2022.
45. Levy, 2022.
46. Levin, 2021.
47. Levin, 2022.
48. Levin, 2022.
49. Rushkoff, 2019, pp. 93–4.

Chapter 10

1. Greenwood, 1961.
2. Greenwood, 1961.
3. Campbell, 2002.
4. Gidley, 2017.
5. Lachman, 2008.
6. Goodrick-Clarke, 2019.
7. Goodrick-Clarke, 2019.
8. Klein, 2023.
9. Beres & Remski, 2021.
10. Lachman, 2008.
11. Gatland & Jefferis, 1979/2023.
12. Gatland & Jefferis 1979/2023, p. 28.
13. Gatland & Jefferis, 1979/2023.
14. Gatland & Jefferis, 1979/2023.
15. Wikipedia Contributors, 2019c.
16. United States Congress Senate Select Committee on Intelligence, 1977.
17. Solon, 2016.
18. Radin, n.d.
19. Carter, 2023.
20. Rushkoff, 2022, p. 113.
21. Rushkoff, 2022, p. 114.
22. Bauer, 2001.
23. Hoffman, 2019.
24. Whitehead, 1925/1967, p. 91.
25. Discussion in Varela, 1997, p. 169.

26. Devereux, 1996, p. 41.
27. Wilson (1986) further discusses the implications of 'tuning in' and 'tuning out' different aspects of reality.
28. Hardy, 1965; Hardy, Harvie, & Koestler, 1974.
29. Hay, 2007, p. 2.
30. Holt, 2008.
31. Hay, 2007, p. 3.
32. Maté & Maté, 2022.
33. Maté & Maté, 2022, p. 452–3.
34. Maté & Maté, 2022, p. 461.
35. Isham, Elf, & Jackson, 2022.
36. Coyne, 2009, p. 244.
37. Holton, 1968.
38. Sheldrake, 2020.
39. Kellehear, 1996, p. 265.
40. Seth et al., 2018.
41. Sheldrake, 2020.
42. Sheldrake, 2020.
43. McGilchrist, 2021.
44. Bendell, 2023, p. 260.
45. Bendell, 2023, p. 374.
46. Tallis (2011) also discusses this problem in the context of Libet's experiments.
47. Bendell, 2023, p. 380.
48. Tegmark (2017) discusses this and other AI-driven, technotopian scenarios.
49. Bendell, 2023.
50. Colborn, 2018.
51. Included in Huxley, 2009b.

Bibliography

Acunzo, D.J. (2023). 'Special Subsection Afterword: ESP Research and Cognitive Neuroscience: Possibly Incompatible – but Methodologically Complementary', *Journal of Scientific Exploration* 37; 3, pp. 536–543.

Albert, D.Z. (2023). 'Philosophy of Physics – Nonlocality', [online] *Britannica*. Available at https://www.britannica.com/topic/philosophy-of-physics/Nonlocality

Alcock, J.E. (2003). 'Give the Null Hypothesis a Chance: Reasons to Remain Doubtful about the Existence of Psi', *Journal of Consciousness Studies* 10; 6–7, pp. 29–50.

Anderson, R. & Braud, W. (2011). *Transforming Self and Others Through Research: Transpersonal Research Methods and Skills for the Human Sciences and Humanities*. Albany: State University of New York Press.

Ankarloo, B. & Clark, S. (1999) (eds.). *Witchcraft and Magic in Europe, Volume 5: The Eighteenth and Nineteenth Centuries*. Philadelphia, PA: University of Pennsylvania Press.

Baars, B. (2017). 'The Global Workspace Theory of Consciousness: Predictions and Results', in Schneider, S. & Velmans, M. (eds). *The Blackwell Companion to Consciousness*. London: Wiley Blackwell, pp. 229–242.

Bauer, H.H. (2001). *Science or Pseudoscience: Magnetic Healing, Psychic Phenomena, and Other Heterodoxies*. Urbana IL: University of Illinois Press.

Bem, D.J. & Honorton, C. (1994). 'Does Psi Exist? Evidence for an Anomalous Process of Information Transfer', *Psychological Bulletin* Vol. 115, No. 1, pp. 4–18.

Bendell, J. (2023). *Breaking Together: A Freedom-Loving Response to Collapse*. Bristol: Good Works.

Bentall, R.P. (2014). 'Hallucinatory Experiences', in Cardeña, E., Lynn, S.J., & Krippner, S. (eds). *Varieties of Anomalous*

Experience: Examining the Scientific Evidence (2nd ed.). Washington, DC: APA, pp. 109–143.

Beres, D. & Remski, M. (2021). 'Bonus Sample: The Question of Consciousness', *Conspirituality*. Available at https://podcasts.apple.com/gb/podcast/conspirituality/id1515827446?i=1000532821055

Berger, R.E. (1989). 'A Critical Examination of the Blackmore Psi Experiments', *Journal of the American Society for Psychical Research* Vol. 83, pp. 123–44.

Blackmore, S.J. (2020). *Seeing Myself: What Out-of-Body Experiences Tell Us About Life, Death and the Mind*. London: Robinson.

Blackmore, S.J. (2010). 'Why I Had to Change My Mind', in Gross, R. *Psychology: The Science Of Mind And Behaviour*. London: Hodder Education, pp 86–7. Available at https://www.susanblackmore.uk/chapters/why-i-had-to-change-my-mind/

Blackmore, S.J. (1996). *In Search of the Light: The Adventures of a Parapsychologist*. Buffalo, Amherst: Prometheus.

Blackmore, S.J. (1987a). 'A Report of a Visit to Carl Sargent's Laboratory', *Journal of the Society for Psychical Research* Vol. 54, pp. 186–98.

Blackmore, S.J. (1982). *Beyond The Body: An Investigation of Out-of-the-Body Experiences*. London: Heinemann.

Blackmore, S.J. & Troscianko, E.T. (2018). *Consciousness: An Introduction* (3rd ed.). London: Routledge.

Blair, L. & Blair, L. (1988). *Ring Of Fire.* London: Bantam Press.

Bor, D. (2012). *The Ravenous Brain: How the New Science of Consciousness Explains Our Insatiable Search for Meaning*. New York: Basic Books.

Bostrom, N. (2011). 'In Defence Of Posthuman Dignity', in Hansell, G.R. & Grassie, W. (eds). *Transhumanism and Its Critics*. Philadelphia, PA: Metanexus, pp. 55–66.

Bostrom, N. (2003). 'The Transhumanist FAQ A General Introduction Version 2.1'. Available at https://nickbostrom.com/views/transhumanist.pdf

Breaking Convention. (2017). 'Psychedelics debate 2017 – the brain and beyond (Dr Robin Carhart-Harris & Dr David Luke)' [video]. Available at https://youtu.be/H4fJXI1hN4k

Bristow, W. (2023). 'Enlightenment', in Zalta, E.N. & Nodelman, U. (eds). *The Stanford Encyclopedia of Philosophy* (Fall 2023 Edition). Available at https://plato.stanford.edu/archives/fall2023/entries/enlightenment/

Broad, C.D. (1953). *Religion, Philosophy and Psychical Research*. London: Harcourt Brace.

Campbell, D. (2002). 'Island of dreams', [online] *Guardian*, 27 April Available at https://www.theguardian.com/books/2002/apr/27/artsandhumanities.highereducation

Cardeña, E. (2018). 'The Experimental Evidence for Parapsychological Phenomena: A Review', *American Psychologist* Vol. 73, No. 5, pp. 663–677. Available at https://doi.org/10.1037/amp0000236

Cardeña, E. (2015). 'Eminent People Interested in Psi', *Psi Encyclopedia*. Available at https://psi-encyclopedia.spr.ac.uk/articles/eminent-people-interested-psi

Cardeña, E., Palmer, J. & Marcusson-Clavertz, D. (2015). *Parapsychology: A Handbook for the 21st Century*. London: McFarland.

Carhart-Harris, R.L., Erritzoe, D., Williams, T., Stone, J.M., Reed, L.J., Colasanti, A., Tyacke, R.J., Leech, R., Malizia, A.L., Murphy, K., Hobden, P., Evans, J., Feilding, A., Wise, R.G., & Nutt, D.J. (2012). 'Neural Correlates of the Psychedelic State as Determined by Fmri Studies with Psilocybin', *Proceedings of the National Academy of Sciences* 109; 6, pp. 2138–2143. Available at https://doi.org/10.1073/pnas.1119598109

Carpenter, J. (2012). *First Sight: ESP and Parapsychology in Everyday Life*. London: Rowman & Littlefield.

Carr, B. (2002). 'Rational Perspectives on the Paranormal: A Review of the Perrott-Warrick Conference Held at Cambridge, 3–5 April 2000', *Journal of Scientific Exploration* 16; 4, pp. 635–650.

Carrington, D. (2024). '2023 smashes record for world's hottest year by huge margin', [online] *Guardian*, 9 January Available at https://www.theguardian.com/environment/2024/jan/09/2023-record-world-hottest-climate-fossil-fuel

Carter, T. (2023). 'Elon Musk says Neuralink can create the "Luke Skywalker solution" amid claims over monkey deaths', [online] *Business Insider*, 21 September. Available at https://www.businessinsider.com/elon-musk-neuralink-create-luke-skywalker-solution-robotic-hands-ai-2023-9?r=US&IR=T

Casad, B.J. & Luebering, J.E. (2023). 'Confirmation Bias', [online] *Encyclopædia Britannica*. Available at https://www.britannica.com/science/confirmation-bias

Castro, M., Burrows, R., & Wooffitt, R. (2014). 'The Paranormal Is (Still) Normal: The Sociological Implications of a Survey of Paranormal Experiences in Great Britain', *Sociological Research Online* Vol. 19, No. 3, pp. 1–15. Available at https://doi.org/10.5153/sro.3355

Chaikin, A. (1994/2019). *A Man on the Moon: The Voyages of the Apollo Astronauts*. London: Penguin.

Chalmers, D.J. (1995). 'Facing up to the problem of consciousness', *Journal of Consciousness Studies* Vol. 2, No. 3, pp. 200–219.

Colborn, M.L.C. (2018). 'Return of the Wild', [online] *Little Blue Marble*, July 27. Available at https://littlebluemarble.ca/2018/07/27/return-of-the-wild/

Conefrey, M. (2012). *Everest 1953: The Epic Story of the First Ascent*. London: Oneworld.

Coyne, J. (2009). *Why Evolution Is True*. Oxford: Oxford Landmark Science.

Cusick, J. (2016). 'Dunblane massacre: Remembering the school shooting 20 years later', [online] *Independent*, 10 March. Available at https://www.independent.co.uk/news/uk/home-news/dunblane-massacre-remembering-the-school-shooting-20-years-later-a6923756.html

Cuthbertson, A. (2023). 'Mind-reading cap turns thoughts into text in world first', [online] *Independent*, 13 December. Available at https://www.independent.co.uk/tech/mind-reading-ai-braingpt-b2463240.html

Dennett, D.C. (1995). *Darwin's Dangerous Idea*. London: Allen Lane.

Descartes, R. (1997). *Philosophical Writings*. London: Workman.

Devereux, P. (1996). *Revisioning The Earth: A Guide to Opening the Healing Channels Between Mind and Nature*. New York: Fireside.

Downing, L. (2021). 'George Berkeley', in Zalta, E.N. (ed.). *The Stanford Encyclopedia of Philosophy* (Fall Edition). Available at https://plato.stanford.edu/archives/fall2021/entries/berkeley/

Duggan, M. (2019). 'Chris Roe', *Psi Encyclopedia*. Available at https://psi-encyclopedia.spr.ac.uk/articles/chris-roe

Edwards, H. (1999). 'Scientists and the paranormal', *The Skeptic* (UK) 12; 3/4, pp. 26–28.

Evans, H. (1982). *Intrusions: Society and the Paranormal*. London: Routledge & Kegan Paul.

Eysenck, H.J. & Sargent, C. (1993). *Explaining the Unexplained: Mysteries of the Paranormal*. London: Prion.

Frankish, K. (2016). *Illusionism as a Theory of Consciousness*. Exeter: Imprint Academic.

French, C.C. (2003). 'Fantastic memories: the relevance of research into eyewitness testimony and false memories for reports of anomalous experiences', *Journal of Consciousness Studies* Vol. 10, Nos. 6–7, pp. 153–174.

French, C.C. & Stone, A. (2014). *Anomalistic Psychology: Exploring Paranormal Belief and Experience*. London: Palgrave MacMillan.

French, C.C. & Wilson, K. (2007). 'Cognitive Factors Underlying Paranormal Beliefs and Experiences', in Della Sala, S. (ed.). *Tall Tales about the Mind and Brain. Separating Fact From Fiction*. Oxford: Oxford University Press, pp. 3–22.

Gardner, H. (1985). *The Mind's New Science*. New York: Basic Books.

Gatland, K. & Jefferis, K. (1979/2023). *The Usborne Book of the Future: A Trip in Time to the Year 2000 and Beyond*. London: Usborne.

Gauld, A. (1968). *The founders of psychical research*. London: Routledge & Kegan Paul.

Gerbic, S. (2013a). 'Wikapediatrician Susan Gerbic discusses her Guerrilla Skepticism on Wikipedia project', [online] *Skeptical Inquirer*, 8 March. Available at https://skepticalinquirer.org/exclusive/wikapediatrician-susan-gerbic-discusses-her-guerrilla-skepticism-on-wikiped/

Gerbic, S. (2013b). 'Egg Balancing and More (Lots More)', [online] *Guerrilla Skepticism on Wikipedia,* 7 December. Available at http://guerrillaskepticismonwikipedia.blogspot.com/2013/12/egg-balancing-and-more-lots-more.html

Gidley, J.M. (2017). *The Future: A Very Short Introduction*. Oxford: Oxford University Press.

Gieryn, T. (1999). *Cultural Boundaries of Science: Credibility on the Line*. Chicago: University of Chicago Press.

Glanvill, J. (1681). *Sadducismus Triumphatus, or, Full and Plain Evidence Concerning Witches and Apparitions: in Two Parts; The First Treating of Their Possibility, the Second of Their Real Existence*. London: J. Collins & S. Lownds. Available at http://name.umdl.umich.edu/A42824.0001.001

Goodrick-Clarke, N. (2019). *The Occult Roots of Nazism: Secret Aryan Cults and Their Influence on Nazi Ideology*. London: Tauris Parke.

Green, J.O. (2018). 'Uri Geller and the reception of parapsychology in the 1970s'. Unpublished Master's Thesis, the University British Columbia, Vancouver.

Greenwood, H.W. (1961). 'Bel-Air Brentwood and Santa Ynez fires worst fire in the history of Los Angeles. Official Report of the Los Angeles Fire Department', [online] lafire.com. Available at https://www.lafire.com/famous_fires/1961-1106_BelAirFire/1961-1106_LAFD-Report_BelAirFire.htm

Gregory, A. (1983). 'Problems in investigating psychokinesis in special subjects'. PhD Thesis, London Metropolitan University.

Gribbin, J. (2012/1984). *In Search of Schrödinger's Cat: Quantum Physics and Reality*. London: Black Swan.

Griffiths, R.R., Richards, W.A., McCann, U., & Jesse, R. (2006). 'Psilocybin Can Occasion Mystical-Type Experiences Having Substantial and Sustained Personal Meaning and Spiritual Significance', *Psychopharmacology* 187; 3, pp. 268–283. Available at https://doi.org/10.1007/s00213-006-0457-5

Grosso, M. (2015). 'The "Transmission" Model of Mind and Body: A Brief History', in Kelly, E.F., Crabtree, A., & Marshall, P. (eds). *Beyond Physicalism: Toward Reconciliation of Science and Spirituality*. London: Rowman & Littlefield, pp. 79–114.

Guerra-Doce, E., Rihuete-Herrada, C., Micó, R., Risch, R., Lull, V., & Niemeyer, H.M. (2023). 'Direct evidence of the use of multiple drugs in Bronze Age Menorca (Western Mediterranean) from human hair analysis', [online] *Scientific Reports* 13(1), p. 4782. Available at https://doi.org/10.1038/s41598-023-31064-2

Gurney, E.F., Myers, F.W.H., & Podmore, F. (1886). *Phantasms of the Living* (2 vols). London: Trübner and co.

Hall, D.F., McFeaters, S.J., & Loftus, E. (1987). 'Alterations in Recollection of Unusual and Unexpected Events', *Journal of Scientific Exploration* 1, No. 1, pp. 3–10.

Hancock, G. (2005). *Supernatural: Meetings with the Ancient Teachers of Mankind*. London: Arrow.

Hansen, G.P. (2001). *The Trickster and the Paranormal*. XLibris.

Hansen, G.P. (1992). 'CSICOP and the Skeptics: An Overview', *Journal of the American Society for Psychical Research* 86, No. 1, pp. 19–63.

Harari, Y.N. (2020). 'Commencement Speech 2020: Congratulations, You Are Now Hackable Animals' [video]. Available at https://youtu.be/rUIwIuyvu8w

Hardy, A. (1965). *The Living Stream: A Restatement of Evolution Theory and Its Relation to the Spirit of Man*. London: Harper & Row.

Hardy, A., Harvie, R., & Koestler, A. (1974). *The Challenge of Chance: A Mass Experiment in Telepathy and Its Unexpected Outcome*. New York: Random House.

Harris, M. (2023). *Palo Alto: A History of California, Capitalism, and the World*. London: Riverrun.

Hartranft, C. (trans). (2003). *The Yoga-Sutra of Patanjali*. Boulder: Shambhala.

Hay, D. (2007). *Why Spirituality Is Difficult for Westerners*. Exeter: Imprint Academic.

Henshilwood, C.S., d'Errico, F., van Niekerk, K.L., Dayet, L., Queffelec, A., & Pollarolo, L. (2018). 'An abstract drawing from the 73,000-year-old levels at Blombos Cave, South Africa', *Nature*, 562; 7725, pp. 115–118. Available at https://doi.org/10.1038/s41586-018-0514-3

Hoffman, D.D. (2019). *The Case Against Reality: How Evolution Hid the Truth From Our Eyes*. London: Allen Lane.

Holt, J. (2008). 'The Laughter of Copernicus', in Broderick, D. (ed.). *Year Million: Science at the Far Edge of Knowledge*. New York: Atlas & Co, pp. 1–20.

Holton, G. (1968). 'Mach, Einstein, and the Search for Reality', *Daedalus* Vol. 97, No. 2, *Historical Population Studies*, pp. 636–673.

Honorton, C. (1993). 'Rhetoric over Substance: The Impoverished State of Skepticism', *Journal of Parapsychology* Vol. 57, pp. 191–214.

Honorton, C. (1985). 'Meta-analysis of ganzfeld research: a response to Hyman', *Journal of Parapsychology* Vol. 49, No. 1, pp. 51–91.

Horgan, J. (2016). 'Flashback: My Report on First Consciousness Powwow in Tucson. How Far Has Science Come Since Then?', *Scientific American*, 26 April [blog]. Available at https://blogs.scientificamerican.com/cross-check/flashback-my-report-on-first-consciousness-powwow-in-tucson-how-far-has-science-come-since-then/

Horowitz, M. (2020). 'The Man Who Destroyed Skepticism', *Skeptical About Skeptics*, 26 October. Available at https://skepticalaboutskeptics.org/investigating-skeptics/whos-who-of-media-skeptics/james-randi/mitch-horowitz-the-man-who-destroyed-skepticism/

Humphrey, N. (1996). *Soul Searching: Human Nature and Supernatural Belief*. London: Vintage.

Hunter, M. (2020). *The Decline of Magic: Britain in the Enlightenment*. London: Yale University Press.

Huxley, A. (2009a). *The Doors of Perception and Heaven and Hell*. London: Harper Perennial.

Huxley, A. (2009b). *The Perennial Philosophy*. London: Harper Perennial.

Huxley, A. (1962/2002). *Island*. London: Harper Perennial.

Huxley, A. (1932/2007). *Brave New World*. London: Vintage.

Hyman, R. (1996). 'Evaluation of a Program on Anomalous Mental Phenomena', *Journal of Scientific Exploration* Vol. 10, No. 1, pp. 31–58.

Hyman, R. (1985). 'The Ganzfeld Psi Experiment: A Critical Appraisal', *Journal of Parapsychology* Vol. 49, No.1, pp. 3–49.

Hyman, R. & Honorton, C. (1986). 'A Joint Communiqué: The Psi Ganzfeld Controversy', *Journal of Parapsychology* Vol. 50, pp. 351–64.

IIT Concerned, Fleming, S. M., Frith, C., Goodale, M., Lau, H., LeDoux, J. E., ... Slagter, H. A. (2023). 'The Integrated Information Theory of Consciousness as Pseudoscience', *PsyArXiv Preprints*, 16 September. Available at https://doi.org/10.31234/osf.io/zsr78

Inglis. B. (1984). *Science and Parascience: A History of the Paranormal*. London: Hodder & Stoughton.

Institute of Noetic Sciences (2016). '"We Are One" — Edgar Mitchell (1930–2016)' [video]. Available at https://youtu.be/R1Kzs9wGlYI?si=7wGZZ9wiKD1gnUrM

Internjack. (2015). 'Precognition', *Parapsychological Association*. Available at https://parapsych.org/articles/53/342/precognition.aspx

Isham, A., Elf, P., & Jackson, T. (2022). 'Self-transcendent experiences as promoters of ecological wellbeing? Exploration of the evidence and hypotheses to be tested', *Frontiers in Psychology* 13. Available at https://doi.org/10.3389/fpsyg.2022.1051478

James, W. (1917/1982). *The Varieties of Religious Experience: A Study in Human Nature*. London: Penguin.

James, W. (1890/2023). 'The Hidden Self', [online] *Wikisource*. Available at https://en.wikisource.org/w/index.php?title=The_Hidden_Self&oldid=3748133

Josephson-Storm, J.A. (2017). *The Myth of Disenchantment: Magic, Modernity and the Birth of the Human Sciences*. London: The University of Chicago Press.

Kastrup, B. (2019). 'Brain image extraction: Is it metaphysically significant?'. Available at https://www.bernardokastrup.com/2019/11/brain-image-extraction-is-it.html

Kastrup, B. (2018). 'The fix is worse than the problem: a reply to psychedelic researchers'. Available at https://www.bernardokastrup.com/2018/10/the-fix-is-worse-than-problem-reply-to.html

Kellehear, A. (1996). *Experiences Near Death: Beyond Medicine and Religion*. Oxford: Oxford University Press.

Kelly, E.F (2022). 'Some conceptual and empirical shortcomings of the Integrated Information Theory of Consciousness (IIT)', *Journal of Anomalous Experience and Cognition* Vol. 2, No. 2, pp. 223–263.

Kelly, E.F. (2015a) 'Empirical challenges of theory construction', in Kelly, E.F., Crabtree, A., & Marshall, P. (eds). *Beyond Physicalism: Toward Reconciliation of Science and Spirituality*. Lanham, Maryland: Rowman & Littlefield, pp. 3–38.

Kelly, E.F. (2015b). 'Toward Reconciliation of Science and Spirituality: A Brief History of the "Sursem" Project', *Edgescience* No. 22, pp. 3– 7.

Kelly, E.F & Kastrup, B. (2018). 'Misreporting and confirmation bias in psychedelic research', [online] *Scientific American Blog Network*. Available at https://blogs.scientificamerican.com/observations/misreporting-and-confirmation-bias-in-psychedelic-research/

Kelly, E.F., Kelly, E.W., Crabtree, A., Gauld, A., Grosso, M., & Greyson, B. (2007). *Irreducible Mind: Toward a Psychology for the 21st Century*. Lanham, Maryland: Rowman & Littlefield.

Kelly, E.F. & Locke, R.G. (2009). *Altered States of Consciousness and Psi: An Historical Survey And Research Prospectus*. New York: Parapsychology Foundation.

Kelly, E.F. & Marshall, P. (2021). *Consciousness Unbound: Liberating Mind from the Tyranny of Materialism*. Lanham, Maryland: Rowman & Littlefield.

Keshmiri, S. (2020). 'Entropy and the Brain: An Overview', *Entropy* 22(9), p. 917. Available at https://doi.org/10.3390/e22090917

Klein, N. (2023). *Doppelganger: A Trip into the Mirror World*. London: Allen Lane.

Krippner, S., & Friedman, H. L. (eds) (2010). *Debating Psychic Experience*. Santa Barbara, CA: Praeger.

Lachman, G. (2008). *Politics and The Occult: The Left, The Right, And The Radically Unseen*. Wheaton, IL: Quest Books.

Leal, W. (2008). 'Bombyx Mori and Bombykol' [video]. Available at https://www.youtube.com/watch?v=6dwy7HcCuVI accessed on 5 October 2019.

Lee, J. (2021). 'From scientists to salesmen: a review of Noah Hutton's *In Silico*', [online] *Science for the People*, 7 October. Available at https://magazine.scienceforthepeople.org/online/from-scientists-to-salesmen/

Lenharo, M. (2023a). 'Philosopher wins consciousness bet with neuroscientist', [online] *Nature* Vol. 619, pp. 14–15.

Lenharo, M. (2023b). 'Consciousness theory slammed as 'pseudoscience' — sparking uproar', [online] *Nature*. Available at https://doi.org/10.1038/d41586-023-02971-1

Levin, S.B. (2022). 'Silicon Valley's Favorite Weird Philosophy Is Fundamentally Wrong', [online] *Slate*. Available at https://slate.com/technology/2022/03/silicon-valley-transhumanism-eugenics-information.html

Levin, S.B. (2021). *Posthuman Bliss?: The Failed Promise of Transhumanism*. New York: Oxford University Press USA.

Levy, R. (2022). 'Musk's Neuralink faces federal inquiry after killing 1,500 animals in testing', [online] *Guardian*, 5 December Available at https://www.theguardian.com/technology/2022/dec/05/neuralink-animal-testing-elon-musk-investigation

Lewis-Williams, D. (2002). *The Mind in the Cave*. London: Thames & Hudson.

Linszen, M.M.J., de Boer, J.N., Schutte, M.J.L., et al. (2022). 'Occurrence and phenomenology of hallucinations in the general population: A large online survey', *Schizophrenia* Vol. 8, No. 41. Available at https://doi.org/10.1038/s41537-022-00229-9

Loftus, E. (1996). *Eyewitness Testimony*. Harvard: Harvard University Press, 1996.

Lorraine. (2018). 'Prof. Brian Cox explains why ghosts aren't real | *Lorraine*' (31 October) [video]. Available at https://www.youtube.com/watch?v=FUbJEUQYjKk&t=7s

Luke, D. (2022). 'Anomalous Psychedelic Experiences: At the Neurochemical Juncture of the Humanistic and Parapsychological', *Journal of Humanistic Psychology* 62(2), pp. 257–297. Available at https://doi.org/10.1177/0022167820917767

Luke, D. (2019). *Otherworlds: Psychedelics and Exceptional Human Experience* (new ed.). London: Aeon Academic.

MacAskill, W. (2023). *What We Owe the Future*. London: Oneworld.

Marshall, P. (2022) 'Does Mystical Experience Give Access To Reality?', *Religions* Vol. 13, No. 10, p. 983. Available at https://doi.org/10.3390/rel13100983

Marshall, P. (2015). 'Mystical Experiences As Windows On Reality', in Kelly, E.F., Crabtree, A, & Marshall, P. (eds). *Beyond physicalism: toward reconciliation of science and spirituality*. London: Roman & Littlefield, pp. 39–76.

Marshall, P. (2005). *Mystical Encounters with the Natural World: Experiences and Explanations*. Oxford: Oxford University Press.

Martin, B. (2023). 'Postmaterialism, Anyone?', *Social Epistemology Review and Reply Collective* Vol. 12, No. 3, pp. 17–26, p. 22. Available at https://wp.me/p1Bfg0-7F5

Martin, B. (2021). 'Policing orthodoxy on Wikipedia: Skeptics in action?' *JCOM* Vol. 20, No. 2, A09. Available at https://doi.org/10.22323/2.20020209

Martin, B. (2018). 'Persistent Bias on Wikipedia: Methods and Responses', *Social Science Computer Review* Vol. 36, No. 3, pp. 379–388. Available at https://doi.org/10.1177/0894439317715434

Marwaha, S.B. (2020a). 'Stargate (Part 1): The US Government Sponsored Psi Research Program', *Paranormal Review* No. 94, pp. 4–13.

Marwaha, S.B. (2020b). 'Stargate (Part 2): The Utility of Informational Psi the Stargate Operational Remote Viewing Program', *Paranormal Review* No. 95, pp. 4–13.

Maté, G. & Maté, D. (2022). *The Myth of Normal: Trauma, Illness & Healing in a Toxic Culture*. London: Vermilion.

Matlock, J.G. (2021). *Signs of Reincarnation: Exploring Beliefs, Cases, and Theory*. Lanham, Maryland: Rowman & Littlefield.

May, E.C. (1996). 'The American Institutes for Research Review of the Department of Defense's STAR GATE Program: A Commentary', *Journal of Scientific Exploration* Vol. 10, No. 1, pp. 89–107.

May, E.C. & Marwaha, S.B. (2014). *Anomalous Cognition: Remote Viewing Research and Theory*. London: McFarland & Co.

May, E.C., Utts, J.M., & Spottiswoode, S.J.P. (1995). 'Decision Augmentation Theory: toward a model of anomalous mental phenomena', *The Journal of Parapsychology* Vol. 59, pp. 195–220.

Mayer, E.L. (2008). *Extraordinary Knowing: Science, Skepticism, and the Inexplicable Powers of the Human Mind*. New York: Random House.

McGilchrist, I. (2021). *The Matter With Things: Our Brains, Our Delusions and the Unmaking of the World* (2 Vols). London: Perspectiva Press.

McGuire, W. (2022). *Hothouse Earth: An Inhabitant's Guide*. London: Icon.

McLuhan, R. (2010). *Randi's Prize: What Skeptics Say About The Paranormal, Why They Are Wrong & Why It Matters*. Leicester: Troubador.

Midgley, M. (2014). *Are You an Illusion*. London: Acumen.

Miller, G.A. (1956). 'The magical number seven, plus or minus two: Some limits on our capacity for processing information', *Psychological Review* Vol. 63, No. 2, pp. 81–97.

Milton, J. & Wiseman, R. (1998). 'Experiment one of the SAIC remote viewing program: A critical re-evaluation', *Journal of Parapsychology* Vol. 62, No. 4, pp. 297–308.

Milton, J. & Wiseman, R. (1999). 'Does psi exist? lack of replication of an anomalous process of information transfer', *Psychological Bulletin* Vol. 125, No. 4, pp. 387–391. Available at https://doi.org/10.1037/0033-2909.125.4.387

Mitchell, E. (1996). *The Way of the Explorer: An Apollo Astronaut's Journey Through the Material and Mystical Worlds*. New York: G.P. Putnam's Sons.

Mobbs, D. & Watt, C. (2011). 'There is nothing paranormal about near-death experiences: how neuroscience can explain seeing bright lights, meeting the dead, or being convinced you are one of them', *Trends Cogn. Sci.*, 15(10), October, pp. 447–9. Available at DOI: 10.1016/j.tics.2011.07.010

Mosley, D.M. (2023). 'Dr Michael Mosley: "Seeing" a ghost is a wake-up call you can mute', [online] *Mail Online*. Available at https://www.dailymail.co.uk/health/article-12154281/DR-MICHAEL-MOSLEY-Seeing-ghost-wake-call-mute.html

Moss, P. (1977). *Ghosts over Britain*. London: David & Charles.

Mroczkowski, J. A., & Malozemoff, A. P. (2019). 'Quantum Misuse in Psychic Literature', *Journal of Near-Death Studies* Vol. 37, pp. 131–154. Available at doi:10.17514/JNDS-2019-37-3-p131-154.

Murphy, M. (1992). *The Future of the Body: Explorations in the Further Evolution of Human Behaviour*. New York: Tarcher Putnam.

Myers, F.W.H. (1903/2001). *Human Personality and Its Survival of Bodily Death*. (Abridged version). Charlottesville, VA: Hampton Roads.

Nahm, M., Rousseau, D., & Greyson, B. (2017). 'Discrepancy Between Cerebral Structure and Cognitive Functioning: A Review', *The Journal of Nervous and Mental Disease* Vol. 205, No. 12, pp. 967–972.

Narby, J. (1999). *The Cosmic Serpent: DNA and the Origins of Knowledge*. London: Weidenfeld & Nicholson.

Nemirovsky, I.E., Popiel, N.J.M., Rudas, J., Caius, M., Naci, L., Schiff, N.D., Owen, A.M., & Soddu, A. (2023). 'An implementation of integrated information theory in resting-state fMRI', *Communications Biology*, 6; 1, pp. 1–14. Available at https://doi.org/10.1038/s42003-023-05063-y

NeuroAI UW Laboratory. (2013). 'Dynamics of olfactory neural codes' [video]. Available at https://www.youtube.com/watch?v=FIRYjVnqBp0

Noakes, R. (2019). *Physics and Psychics: The Occult and the Sciences in Modern Britain*. Cambridge: Cambridge University Press.

Nosta, J. (2023). 'The 5th Industrial Revolution: The Dawn of the Cognitive Age', [online] *Psychology Today United Kingdom*. Available at https://www.psychologytoday.com/gb/blog/the-digital-self/202310/the-5th-industrial-revolution-the-dawn-of-the-cognitive-age

Otgaar, H., Howe, M.L., Muris, P., & Merckelbach, H. (2019) 'Dealing With False Memories in Children and Adults: Recommendations for the Legal Arena', *Policy Insights from the Behavioral and Brain Sciences* Vol. 6, No. 1, pp. 87–93. Available at https://doi.org/10.1177/2372732218818584

Paolo Fusar-Poli, Andrés Estradé, Stanghellini, G., Cecilia Maria Esposito, René Rosfort, Mancini, M., Norman, P., Cullen, J.M., Miracle Ayomikun Adesina, Gema Benavides Jimenez, Caroline, Drah, E.A., Julien, M.-H., Lamba, M., Mutura, E.M., Prawira, B., Agus Sugianto, Jaleta Teressa, White, L.A., & Damiani, S. (2023). 'The lived experience of depression: a bottom-up review co-written by experts by experience and academics', *World Psychiatry* Vol. 22, No. 3, pp. 352–365. Available at https://doi.org/10.1002/wps.21111

Parapsychological Association. (2023). 'First sight: a psychological theory of psi', James Carpenter (24 July) [video]. Available at https://youtu.be/wiCyjr_kWgA?si=omAVC3yOY1RNwcIo

Parker, A. (2017). 'Ganzfeld ESP', [online] *Psi Encyclopedia*. Available at https://psi-encyclopedia.spr.ac.uk/articles/ganzfeld-esp

Petersen, R. (2019). 'Taking mushrooms for depression cured me of my atheism', *The Outline*. Available at https://theoutline.com/post/7367/taking-mushrooms-for-depression-cured-me-of-my-atheism

Pilkington, R. (2013). *Men and Women of Parapsychology: Personal Reflections*. Charlottesville, Virginia: Anomalist Books.

Playfair, G. L. (2015). 'Uri Geller', *Psi Encyclopedia*. Available at https://psi-encyclopedia.spr.ac.uk/articles/uri-geller

Playfair, G.L. (2011). *This House Is Haunted: The Amazing Story of the Enfield Poltergeist*. Guildford: White Crow Books.

Pollan, M. (2019). *How to Change Your Mind: The New Science of Psychedelics*. London: Penguin.

Presti, D.E. (2017). 'Altered States of Consciousness: Drug-Induced States', in Schneider, S. & Velmans, M. (eds). *The Blackwell Companion to Consciousness* (2nd ed.). Chichester, UK: John Wiley, pp. 171–186.

Prince, W.F. (1930). *The Enchanted Boundary: Being a Survey of Negative Reactions to Claims of Psychic Phenomena, 1820–1930*. Boston: Boston Society for Psychic Research.

Puharich, A. (1974). *Uri: The Authorised Biography of Uri Geller, the World's Most Famous Psychic*. London: Futura.

Quaglia, S. (2022). 'The Science Behind Why People See Ghosts and Demons', *The Daily Beast* [blog]. Available at https://www.yahoo.com/now/science-behind-why-people-see-004341225.html?guccounter=1

Rabeyron, T. (2020). 'Why Most Research Findings About Psi Are False: The Replicability Crisis, the Psi Paradox and the Myth of Sisyphus', *Frontiers in Psychology* Vol. 11. Available at https://www.frontiersin.org/articles/10.3389/fpsyg.2020.562992/full

Radin, D. (2019). 'Don't look at my hand: A response to "Quantum misuse in psychic literature"', *Journal of Near-Death Studies* Vol. 37, No. 3, pp. 171–3.

Radin, D. (1997). *The Conscious Universe: The Scientific Truth of Psychic Phenomena*. San Francisco, CA: Harper San Francisco.

Radin, D. (n.d.). 'IONS: Applications of psi', *IONS* website. Available at https://noetic.org/wp-content/uploads/2020/12/IONSx-Applications-of-Psi.pdf

Rainey, S., Martin, S., Christen, A., Mégevand, P., Fourneret, E. (2020) 'Brain recording, mind-reading, and neurotechnology: ethical issues from consumer devices to brain-based speech decoding', *Science and Engineering Ethics* Vol. 26, pp. 2295–2311.

Read, R. (2022). *Why Climate Breakdown Matters (Why Philosophy Matters)*. London: Bloomsbury Academic.

Reber, A. S., & Alcock, J. E. (2020). 'Searching for the impossible: parapsychology's elusive quest', *American Psychologist* Vol. 75, No. 3, pp. 391–399.

Reed, C. (2008). 'Dr Albert Hofmann', [online] *Guardian*, 30 April Available at https://www.theguardian.com/science/2008/apr/30/drugs.chemistry

Reincarnate Lama Returns to Thame! (2015) (12 August). Available at http://www.bergadventures.com/blog/2015-08-12/1855/

Rifkin, J. (1999). *Biotech Century: Harnessing the Gene and Remaking the World*. New York: Jeremy Tarcher.

Ring of Fire: An Indonesian Odyssey. (1988). *BBC 2* (13 June–18 July).

Riskin, J. (2016). *The Restless Clock: A History of the Centuries-Long Argument over What Makes Living Things Tick*. Chicago: University of Chicago Press.

Ritchie, S. L. (2021). 'Panpsychism and Spiritual Flourishing: Constructive Engagement with the New Science of

Psychedelics', *Journal of Consciousness Studies* Vol. 28, pp. 268–88.

Rivas, T., Dirven, A., & Smit, R.H. (2016). *The Self Does Not Die: Verified Paranormal Phenomena from Near-Death Experiences*. International Association for Near-Death Studies.

Robbins, R.H. (1959). *The Encyclopaedia of Witchcraft and Demonology*. New York: Spring Books.

Roe, C. (2024). 'Psi and Altered States of Consciousness. SPR Presidential Address', [Lecture] *The Royal Society*. 9 April.

Roe, C. (2021). 'Blackmore-Sargent Controversy — A Reconsideration', *Psi Encyclopedia*. Available at https://psi-encyclopedia.spr.ac.uk/articles/blackmore-sargent-controversy-%E2%80%93-reconsideration

Roe, C. (2019). 'The Value of Spontaneous Experiences', *Paranormal Review* No. 89, pp. 4–5.

Roe, C. (2018). 'Arguing the Case for Parapsychological Research', *Paranormal Review* No. 88, pp. 6–7.

Roe, C. (2017). 'PA Presidential Address 2017: Withering Skepticism', *Journal of Parapsychology* Vol. 81, No. 2, pp 143–159.

Roe, C. & Linnett, R. (2017). 'Content Analysis of Spontaneous Cases of Psi Included in the Alister Hardy Religious Experience Research Centre Database', in Cooper, C. E. (ed.). *60th Annual Convention of the Parapsychological Association: Abstracts of Presented Papers*. Athens, Greece: Parapsychological Association.

Roe, C.A., Holt, N.J., & Simmonds C.A. (2003). 'Considering the Sender as a PK Agent in Ganzfeld ESP Studies', *Journal of Parapsychology* Vol. 67, pp. 129–145.

Rosenthal, R. & Jacobson, L. (1968). 'Pygmalion in the Classroom', *The Urban Review* Vol. 3, No. 1, pp. 16–20.

Royal Society. (2015). 'History of the Royal Society', Royalsociety.org. Available at https://royalsociety.org/about-us/history/

Rushkoff, D. (2022). *Survival of the Richest: Escape Fantasies of the Tech Billionaires*. London: Scribe.

Rushkoff, D. (2019). *Team Human*. New York: W.W. Norton & Company.

Sacks, O. (2012). *Hallucinations*. London: Picador.

Sagan, C. (1996). *The Demon-Haunted World: Science as a Candle in the Dark*. London: Hodder Headline.

Sargent, C. (1987). 'Sceptical Fairytales from Bristol', *Journal of the Society for Psychical Research* Vol. 54, pp. 208–18.

Sawyer, D. (2022). 'What Is the "Unitive Mystical Experience" Triggered by Psychedelic Medicines an Experience of? An Exploration of Aldous Huxley's Viewpoint in Light of Current Data', *Religions* Vol. 13, No. 11, p. 1061. Available at https://doi.org/10.3390/rel13111061

Schurger, A. & Graziano, M. S. A. (2022) 'Consciousness explained or described?', *Neuroscience of Consciousness* Vol. 7, pp. 1–9. Available at https://doi.org/10.1093/nc/niac001

Schwartz, S. (2017). 'Remote Viewing', *Psi Encyclopedia*. Available at https://psi-encyclopedia.spr.ac.uk/articles/remote-viewing

Searle, J. (1997). *The Mystery of Consciousness*. London: Granta Books.

Seth, A. (2021). *Being You: A New Science of Consciousness*. London: Faber & Faber.

Seth, A. & Bayne, T. (2022). 'Theories of Consciousness', *Nature Reviews Neuroscience* Vol. 23, pp. 439–452. Available at https://doi.org/10.1038/ s41583-022-00587-4

Seth, A., Schartner, M., Tagliazucchi, E., Muthukumaraswamy, S., Carhart-Harris, R., & Barrett, A. (2018). 'What Psychedelic Research Can and Cannot Tell Us about Consciousness', *Scientific American Blog Network*. Available at https://blogs.scientificamerican.com/observations/what-psychedelic-research-can-and-cannot-tell-us-about-consciousness/

Shanon, B. *The Antipodes of the Mind: Charting the Phenomena of the Ayahuasca Experience*. Oxford: Oxford University Press, 2002.

Sheldrake, R. (2020). *The Science Delusion* (2nd ed.). London: Coronet.

Simmonds-Moore, C. A & Holt, N. (2007). 'Trait, state and psi: an exploration of the interaction between individual differences, state preference and psi performance in the ganzfeld and waking ESP control', *Journal of the Society for Psychical Research* Vol. 71, pp. 197–215.

Solon, O. (2016). 'Under pressure, Silicon Valley workers turn to LSD Microdosing', *Wired UK*. Available at https://www.wired.co.uk/article/lsd-microdosing-drugs-silicon-valley

Sommer, A. (2019). 'Inspirations from Swedenborg and Feyerabend: A Conversation with Nuclear Physicist Ian J. Thompson', *Forbidden Histories* [blog], 11 May. Available at https://www.forbiddenhistories.com/2019/05/swedenborg-feyerabend-thompson/

Sommer, A. (2016). 'Are you afraid of the dark? Notes on the psychology of belief in histories of science and the occult', *European Journal of Psychotherapy & Counselling* Vol. 18, No. 2, pp. 105–122. Available at https://doi.org/10.1080/13642537.2016.1170062

Sommer, A. (2013). 'Emil du Bois Reymond: Science, Progress and Superstition. An Interview with Gabriel Finkelstein', *Forbidden Histories* [blog], 2 December. Available at https://www.forbiddenhistories.com/2013/12/finkelstein-interview/

Stapp, H. (2009). 'Nonlocality', in Greenberger, D., Hentschel, K., Weinert, F. (eds). *Compendium of Quantum Physics*. Heidelberg: Springer, Berlin, pp. 405–410.

Storm, L., Tressoldi, P.E., & Di Risio, L. (2010). 'Meta-analysis of free-response studies, 1992–2008: Assessing the noise reduction model in parapsychology', *Psychological Bulletin* Vol. 136, No. 4, pp. 471–85.

Strawson, G. (2018). 'The Consciousness Deniers', *New York Review of Books*, 13 March. Available at https://www.nybooks.com/daily/2018/03/13/the-consciousness-deniers/

Tallis, R. (2011). *Aping Mankind: Neuromania, Darwinitis and the Misrepresentation of Humanity*. London: Acumen.

Tang, J., LeBel, A., Jain, S., & Huth, A.G. (2023). 'Semantic reconstruction of continuous language from brain recordings', *Nature Neuroscience* Vol. 26, pp. 858–866.

Targ, R. & Puthoff, H.E. (1977/2005). *Mind-Reach: Scientists Look at Psychical Research*. Charlottesville, VA: Hampton Roads.

Targ, R. & Puthoff, H.E. (1974). 'Information Transmission Under Conditions of Sensory Shielding', *Nature* Vol. 251, 18 October, pp. 602–607.

Tart, C. (1990). *Altered States of Consciousness*. New York: Harper.

Taylor, S. (2023). *DisConnected: The Roots of Human Cruelty and How Connection Can Heal the World*. London: Iff books.

TED2017. (2017). 'Your brain hallucinates your conscious reality' [video]. Available at https://youtu.be/lyu7v7nWzfo

Tegmark, M. (2017). *Life 3.0*. London: Penguin.

Timmermann, C., Roseman, L., Haridas, S., Rosas, F.E., Luan, L., Kettner, H., Martell, J., Erritzoe, D., Tagliazucchi, E., Pallavicini, C., Girn, M., Alamia, A., Leech, R., Nutt, D.J., & Carhart-Harris, R.L. (2022). 'Human brain effects of DMT assessed via EEG-fMRI', *Proceedings of the National Academy of Sciences* Vol. 120, No. 13. Available at https://doi.org/10.1073/pnas.2218949120

Timmermann, C., Roseman, L., Schartner, M., Milliere, R., Williams, L.T.J., Erritzoe, D., Muthukumaraswamy, S., Ashton, M., Bendrioua, A., Kaur, O., Turton, S., Nour, M.M., Day, C.M., Leech, R., Nutt, D.J., & Carhart-Harris, R.L. (2019). 'Neural correlates of the DMT experience assessed with multivariate EEG', *Scientific Reports* 9; 1, pp. 1–13. Available at https://doi.org/10.1038/s41598-019-51974-4.

Tonini, G. (2017a). 'The Integrated Information Theory of Consciousness: An Outline', in Schneider, S. & Velmans, M. (eds). *The Blackwell Introduction to Consciousness*. London: Wiley Blackwell, pp. 243–256.

Tonini, G. (2017b). 'Integrated Information Theory of Consciousness : Some Ontological Considerations', in Schneider, S. & Velmans, M. (eds). *The Blackwell Introduction to Consciousness*. London: Wiley Blackwell, pp. 621–633.

Torres, É. (2022). 'Understanding "longtermism": Why this suddenly influential philosophy is so toxic', *Salon*. Available at https://www.salon.com/2022/08/20/understanding-longtermism-why-this-suddenly-influential-philosophy-is-so/

Torres, É. (2021). 'Why longtermism is the world's most dangerous secular credo', Aeon Essays. *Aeon*. Available at https://aeon.co/essays/why-longtermism-is-the-worlds-most-dangerous-secular-credo

Toulmin, S. (1992). *Cosmopolis: The Hidden Agenda of Modernity*. Chicago: University of Chicago Press.

Tressoldi, P.E. (2011). 'Extraordinary Claims Require Extraordinary Evidence: The Case of Non-Local Perception, a Classical and Bayesian Review of Evidences', *Frontiers in Psychology* Vol. 2. Available at https://doi.org/10.3389/fpsyg.2011.00117

Tressoldi, P.E. & Katz, D. (2023). 'Remote viewing: a 1974–2022 systematic review and meta-analysis', *Journal of Scientific Exploration* Vol. 37, No. 3, pp. 467–489. Available at https://doi.org/10.31275/20232931

Turnbull, D. (1997). 'Reframing science and other local knowledge traditions', *Futures* 29; 6, pp. 551–562. Available at https://doi.org/10.1016/s0016-3287(97)00030-x

Underhill, E. (2002/1930). *Mysticism: A Study in the Nature and Development of Spiritual Consciousness*. New York: Dover.

United States Congress Senate Select Committee on Intelligence (1977). 'Project MKUProject MKULTRA, the

CIA's Program of Research in Behavioral Modification', U.S. Government Printing Office. Available at https://books.google.co.uk/books?id=TEqhqtrF3XEC&pg=PA70&redir_esc=y#v=onepage&q&f=false

Utts, J. (1999). 'The Significance of Statistics in Mind-Matter Research', *Journal of Scientific Exploration* Vol. 13, No. 4, pp. 615–38.

Utts, J. (1995). In Mumford, M.D., Rose, A.M, & Goslin, D.A. (eds). *An Evaluation of Remote Viewing: Research and Applications*. American Institutes for Research.

Varela, F.J. (2002). *Sleeping, Dreaming, and Dying: An Exploration of Consciousness*. Boston: Wisdom Publications.

Vernon, D. (2021). *Dark Cognition: Evidence for Psi and its Implications for Consciousness*. London: Routledge.

Walach, H. (2019). *Beyond a Materialist Worldview: Towards an Expanded Science* (Galileo Commission Report). Galileo Commission. Available at https://galileocommission.org/report/

Walker, M. (2011). 'Ship of Fools: Why Transhumanism Is the Best Bet to Prevent the Extinction of Civilization', in Hansell, G.R. & Grassie, W. (eds.) *Transhumanism and its Critics*. Philadelphia, PA, Metanexus, pp. 94–110.

Watson, J. (1913). 'Psychology as the behaviorist views it', *Psychological Review* Vol. 20, pp. 158–177. Available at https://psychclassics.yorku.ca/Watson/views.htm

Watt, C. & Tierney, I. (2014). 'Psi-related experiences', in Cardeña, E., Lynn, S.J., & Krippner, S. *Varieties Of Anomalous Experience: Examining the Scientific Evidence* (2nd ed.). Washington, DC: APA, pp. 241–272.

Watt, C., Ashley, N., Gillett, J., Halewood, M., & Hanson, R. (2014). 'Psychological factors in precognitive dream experiences: the role of paranormal belief, selective recall and propensity to find correspondences', *International Journal of Dream Research* Vol. 7, No. 1, pp. 1–8.

Wehrstein, K.M. & Duggan, M. (2020). 'Susan Blackmore', *Psi Encyclopedia*. Available at https://psi-encyclopedia.spr.ac.uk/articles/susan-blackmor

Wehrstein, K.M. & McLuhan, R. (2020). 'James Randi', *Psi Encyclopedia*. Available at https://psi-encyclopedia.spr.ac.uk/articles/james-randi

Weinberg, S. (1977). *The First Three Minutes*. London: André Deusch.

Weiler, C. (2020). *Psi Wars: TED, Wikipedia and the Battle for the Internet*. London: White Crow.

Westlin, C., Theriault, J. E., Katsumi, Y., Nieto-Castanon, A., Kucyi, A., Ruf, S. F., . . ., & Barrett, L.F. (2023). 'Improving the study of brain-behavior relationships by revisiting basic assumptions', *Trends in Cognitive Sciences*. Published online 2 February 2023. Available at https://doi.org/10.1016/j.tics.2022.12.015

Westphal, J. (2016). *The Mind–Body Problem*. London: MIT Press.

Westrum, R. (2011). 'Hidden Events and Closed Minds: The Case of Battered Children', *Edgescience* No. 8, pp. 10–12. Available at https://www.scientificexploration.org/edgescience/8

White, C. (2014). 'The science delusion', interview by Linda Heuman, *Tricycle*, Spring 2014, pp. 72–77, pp. 108–109.

White, M. (1998). *Isaac Newton: The Last Sorcerer*. London: Fourth Estate.

White, R.A. (1999). 'Exceptional Human Experiences: A Brief Overview', *Exceptional Human Experiences Network*. Available at https://www.ehe.org/display/ehe-page53e5.html?ID=5

The White House (2013). 'Fact Sheet: BRAIN Initiative', Office of the press Secretary the White House President Barack Obama, 2 April. Available at https://obamawhitehouse.archives.gov/the-press-office/2013/04/02/fact-sheet-brain-initiative

Whitehead, A.N. (1925/1967). *Science and the Modern World*. London: Free Press.

Whitehead, C. (2004). 'Everything I Believe. Might Be a Delusion. Whoa! Tucson 2004: Ten years on, and are we any nearer to a Science of Consciousness?', *Journal of Consciousness Studies* Vol. 11, No. 12, pp. 68–88.

Wikipedia Contributors. (2023). 'List of psychedelic drugs', [online] *Wikipedia*. Available at https://en.wikipedia.org/wiki/List_of_psychedelic_drugs

Wikipedia Contributors. (2020a). 'Hungerford massacre', [online] *Wikipedia*. Available at https://en.wikipedia.org/wiki/Hungerford_massacre

Wikipedia Contributors. (2020b). 'Parapsychology', [online] *Wikipedia*. Available at https://en.wikipedia.org/wiki/Parapsychology

Wikipedia Contributors. (2020c). 'Rationalisation (sociology)', [online] *Wikipedia*. Available at https://en.wikipedia.org/wiki/Rationalization_(sociology)

Wikipedia Contributors. (2019a). '2004 Indian Ocean earthquake and tsunami', [online] *Wikipedia*. Available at https://en.wikipedia.org/wiki/2004_Indian_Ocean_earthquake_and_tsunami

Wikipedia Contributors. (2019b). 'Dunblane massacre', [online] *Wikipedia*. Available at https://en.wikipedia.org/wiki/Dunblane_massacre

Wikipedia Contributors. (2019c). 'Electricity', [online] *Wikipedia*. Available at https://en.wikipedia.org/wiki/Electricity

Wikipedia Contributors. (2019d). 'Modernism', [online] *Wikipedia*. Available at https://en.wikipedia.org/wiki/Modernism

Wikipedia Contributors. (2019e). 'Pan Am Flight 103', [online] *Wikipedia*. Available at https://en.wikipedia.org/wiki/Pan_Am_Flight_103

Williams, B.J. (2011). 'Revisiting the Ganzfeld ESP Debate: A Basic Review and Assessment', *Journal of Scientific Exploration* Vol. 25, No. 4, pp. 639–61.

Willin, M. (2019). *The Enfield Poltergeist Tapes*. Guildford: White Crow Books.

Wilson, R.A. (1986). *The New Inquisition*. Tempe, Arizona: New Falcon Publications.

Winter, N. R., Blanke, J., Leenings, R., Ernsting, J., Fisch, L., Sarink, K., ..., & Hahn, T. (2023). 'A Systematic Evaluation of Machine Learning-Based Biomarkers for Major Depressive Disorder Across Modalities', medRxiv.org. Available at https://doi.org/.10.1101/2023.02.27.23286311

World Economic Forum. (2023). 'Davos AM23 — Ready for Brain Transparency? — English' [video]. Available at https://www.weforum.org/videos/davos-am23-ready-for-brain-transparency-english

Wyatt, T.D. (2017). *Animal Behaviour: A Very Short Introduction*. Oxford: Oxford University Press.

Young, C.H. (2023). *Spinning Out: Climate Change, Mental Health and Fighting for a Better Future*. London: Footnote Press.

Zingrone, N.L. (n.d.). 'On the Critics of Parapsychology', *Skeptical About Skeptics*. Available at https://skepticalaboutskeptics.org/examining-skeptics/nancy-l-zingrone-james-alcock-fails-to-go-the-distance

Author Biography

Matt Colborn is an author, lecturer, broadcaster, and artist. He currently teaches consciousness studies on the Alef Trust Master's programme. Matt holds an MSc with distinction in cognitive science from the University of Birmingham and a DPhil in biology from the University of Sussex. His first non-fiction book, *Pluralism and the Mind,* was published in 2011, and his first collection of short stories, *City in the Dusk,* in 2013. He has been a member of the Society for Psychical Research since 1998, which he joined because of a lifelong fascination with the paranormal. He also loves trekking and climbing, and in 2023 made it to Everest base camp.

Note to Reader

From the Author: Thank you for purchasing *What Lies Beyond*. My sincere hope is that you enjoyed reading this book! If you have a few moments, please feel free to add your review of the book to your favourite online site. Also, if you would like news, updates, and additional material, please visit my website or the *What Lies Beyond* book blog on Substack. Sincerely, Matt Colborn.

Website: https://www.drmattcolborn.co.uk

What Lies Beyond Substack blog: https://whatliesbeyond.substack.com

ESSENTIA

Essentia Books, a collaboration between Collective Ink Ltd. and Essentia Foundation, publishes rigorous scholarly work relevant to metaphysical idealism, the notion that reality is essentially mental in nature. For more information on modern idealism, please visit www.essentiafoundation.org.